Investigating the Supernatural

Investigating the Supernatural

From Spiritism and Occultism to Psychical Research and Metapsychics in France, 1853–1931

SOFIE LACHAPELLE

The Johns Hopkins University Press
Baltimore

© 2011 The Johns Hopkins University Press
All rights reserved. Published 2011
Printed in the United States of America on acid-free paper
2   4   6   8   9   7   5   3   1

The Johns Hopkins University Press
2715 North Charles Street
Baltimore, Maryland 21218-4363
www.press.jhu.edu

Library of Congress Cataloging-in-Publication Data
Lachapelle, Sofie.
Investigating the supernatural : from spiritism and occultism to psychical
research and metapsychics in France, 1853–1931 / Sofie Lachapelle.
       p.   cm.
Includes bibliographical references (p.   ) and index.
ISBN-13: 978-1-4214-0013-6 (hardcover : alk. paper)
ISBN-10: 1-4214-0013-8 (hardcover : alk. paper)
1. Occultism—France—History—19th century. 2. Occultism—France—
History—20th century. 3. Supernatural—France—History—19th century.
4. Supernatural—France—History—20th century. 5. Parapsychology—
France—History—19th century. 6. Parapsychology—France—History—
20th century. I. Title.
BF1434.F8L33   2011
130.944'09034—dc22   2010042496

A catalog record for this book is available from the British Library.

Special discounts are available for bulk purchases of this book.
For more information, please contact Special Sales at 410-516-6936
or specialsales@press.jhu.edu.

The Johns Hopkins University Press uses environmentally friendly book
materials, including recycled text paper that is composed of at least
30 percent post-consumer waste, whenever possible. All of our book
papers are acid-free, and our jackets and covers are printed on paper
with recycled content.

À Ariane

Contents

Investigating the Supernatural

Introduction

Between 1859 and 1862, the popular science writer Louis Figuier, a former professor at the Paris École de pharmacie, published his *Histoire du merveilleux dans les temps modernes*, a momentous work in four volumes relating many mysterious phenomena witnessed throughout the eighteenth and nineteenth centuries. Featured in his collection of supernatural occurrences were divining rods, pendulums, cases of group possessions, electric girls, and mediums communicating with the dead. By the late 1850s, Figuier was just beginning what would become a very successful career as a popularizer of science. Through his books on the latest scientific and technological developments, he would continue to introduce his contemporary readers to the wonders of the modern world and the marvels of human advances. Following the death of his son in 1870, Figuier, who had already shown an interest in the supernatural, lunged even further into the subject and began to write on the soul and its survival after death. References to life and immortality found their way into his popular writings on science, particularly his work in astronomy and biology. For him, life was central to the universe. It existed throughout the solar system, he believed. Souls migrated from one planet to another, with the sun as their final destination. It was only a matter of time, he speculated, before science would provide proof of all of this and succeed in incorporating spiritual concerns and supernatural occurrences into its corpus. Such was the necessary way of progress.[1]

Figuier's optimism regarding science was not unusual; neither was the presence of spiritual interests in his popular accounts of scientific advancements. During the nineteenth century, numerous technological innovations and scientific advances transformed the consciousness and experiences of Europeans. Across the continent, these advances and novelties blurred the lines between the natural and the supernatural. Human inventions seemed to be reaching into the realm of the fantastic. Telegraphs, photographs, and trains, to name a few, had brought about changes that would have appeared impossible to previous generations. Popularizers of science played with the sense of wonder that recent inventions inspired. In Figuier's writing, for example, the latest innovations were described as amazing marvels, ready to enter daily life. Science and technology created enchantment; they made the magical seem possible.

All these developments impacted the Catholic Church, which simultaneously experienced internal difficulties and a sense of resurgence in the second half of the nineteenth century. On the official front, both Pius IX (1846–78) and Leo XIII (1878–1903) faced numerous challenges. For one thing, the unification of Italy considerably reduced the papacy's material power, leaving the pope in charge of a shrunken territory within the city of Rome. In a series of measures adopted in response to the growing influence of secular trends across Europe, the Vatican proclaimed first the dogma of the Immaculate Conception in 1854 and then the dogma of Papal Infallibility in 1870, following the First Vatican Council.

While ecclesiastics worried about the Church's diminished influence, this did not affect the large majority of the population, for whom Catholicism remained a vibrant set of beliefs and practices, as evidenced by a growing interest in the more tangible experiences of spirituality. Tales of visionaries, stigmatics, demonics, and other believers experiencing physical manifestations of their faith inspired sensational books and pamphlets, pilgrimages, and claims of miracle cures in believers. At Lourdes, the young Bernadette Soubirous' visions of the Virgin Mary in 1858 led to the building of a sanctuary and the development of a commercially successful pilgrimage tradition, which continues to this day. Late nineteenth-century France also witnessed a spiritual revival in urban centers, apparent in the construction of new sites of worship, the most important of which was the imposing Basilique du Sacré-Coeur on the butte Montmartre in Paris. This was a world in which the supernatural was being constructed and concretized, the Virgin Mary directly addressed some of her devotees, claims of miracle cures at sanctuaries abounded, communities found themselves in the grasp of epidemics of demonic possession, and the dead communicated with the living at séances.[2]

These various tangible experiences of a spiritual kind took place amid developments in medicine and psychiatry and the establishment of the new discipline of psychology. Accounts by alienists and psychiatrists of hysteria, suggestion, and hypnotism led physicians to a sphere that had traditionally been the purview of priests and religious orders. In 1883, the intrusion was formalized with the creation the Bureau of Medical Consultations at the Sanctuary of Lourdes. This meant that every claimed miracle associated with the sanctuary would have to be investigated by physicians on the site. In 1892, Jean-Martin Charcot, the famous neurologist at the Salpêtrière hospital, published "La foi qui guérit," an article in which he introduced the concept of faith healing to explain the miraculous recoveries witnessed at Lourdes. He argued that such cures could only occur when the disease was hysterical in origin and the patient was particularly suggestible. Charcot was not alone. Stories of mystics fill the pages of psychiatric works of the period. Examples range from the antagonism of the neurologist Désiré-Magloire Bourneville and his Bibliothèque diabolique, a series in which each volume reexamined a previous or present case of possession or other religious manifestation in pathological terms, to the more sympathetic stance of the psychologist Pierre Janet, who hoped to understand possession, ecstasy, and stigmata in physiological terms. By the end of the nineteenth century, the medical and human sciences had fully infiltrated the religious sphere.[3]

In the sciences more broadly, the period was marked by a growth of professionalization. The borders between scientists and laymen were becoming clearer. Increasingly, science became an activity practiced by a distinct, specially trained group of individuals in a specific set of spaces. In France, this professionalization was particularly marked and took place earlier in the century than anywhere else. As research establishments and universities in Paris and around the country provided positions for those with the necessary skills and training, science became a career. At the center of it all, the Académie des sciences, an institution controlled by its one hundred and fifty members (three times the number of members of the Royal Society in Britain), gave the nation a strong direction in scientific development and research through recognition, financial support, contests, and prizes.[4]

The century also saw the emergence of an industry of popular science, no doubt spurred on by the expansion of the middle class and the growing demand for leisure activities and self-improvement. Scientific lectures, museum exhibits, and popular science books allowed middle-class men, women, and children to find entertainment in discovering and enjoying scientific wonders without needing to fully grasp the concepts and theories behind them. For the general public, science was made edifying, comprehensible, marvelous, and spectacular all at once.[5] Many popularizers like Figuier incorporated claims of the supernatural into their popular accounts, presenting them as marvels on par with electricity or chemistry, soon to be integrated into the scientific corpus. They emphasized the work of scientists and others seeking to gain an understanding of these phenomena.

Of course, there was no consensus as to what counted as a supernatural or a natural occurrence and what did not, but supernatural phenomena were generally understood to be events or experiences that apparently transcended the laws of nature. They were usually attributed to powers that either violated or went beyond natural forces, and they tended to be unpredictable and difficult to observe, control, and reproduce, but often spectacular to witness. This made them challenging if not impossible to investigate through experimental methods. For those interested in such phenomena, several different explanations were possible. For some, supernatural events and experiences were clearly fraudulent, the product of trickery and illusion. For others, they had natural causes that were yet to be discovered by science. If there was no agreement on which phenomena were real and how to explain them, a significant number of individuals and groups, scientists among them, were certainly interested in their investigation.

This book tells the story of the various attempts to explore the supernatural in France between the 1850s and the 1930s using the phenomena witnessed at séances as a connecting thread. Five different groups are discussed in particular: (1) the spiritists, who looked for a connection between the living and the dead at séances, from which they hoped a science of the spiritual realm and the afterlife would emerge; (2) the occultists, who sought to connect ancient wisdom and revelations with contemporary science to develop a "complete" science; (3) the physicians, psychiatrists, and psychologists who regarded claims of supernatural experience as pathological; (4) the psychical researchers, who invited the public to participate in the development of a new field of research by sharing their personal experiences of unexplained phenomena; and (5) the *métapsychistes*, or metapsychists, who believed that psychical phenomena were the key to the development of an entirely new science. While they all considered seemingly supernatural phenomena worthy of study, each of these groups differed in its approach to the investigation of the supernatural and had a distinct understanding of what was meant by scientific exploration and explanation. Their various interactions with the sciences, both successful and unsuccessful, inform the complex nature of the relationship between French science and the supernatural at the time. They point to both the promises and the problems associated with attempts to bring unexplained wonders into the sciences, as well as the limits of science itself when wrestling with such unorthodox practices and knowledge.

Investigations of the supernatural were not new, of course, but throughout the second half of the nineteenth century, they became more visible as parts of several discourses present in and around the sciences of the period. In France, the 1850s saw the appearance of two movements around which these investigations would converge; spiritism, which developed around séances, and modern occultism. In 1853, the French became fascinated with turning tables and séances. Initially introduced as a diversion, the practice soon aroused the interest of religious and scientific authorities alike. In 1857, Allan Kardec presented spiritism as a religious doctrine based on revelations by spirits during séances. Spiritists believed that séances would lead to a revolution in both science and religion; that they would bring about the dawn of a spiritual science and a faith supported by concrete evidence. For them, the scientific and the religious spheres were not to be seen as opposites; hope lay not in one or in the other, but in a reconciliation of the two.

While spiritists were exploring the otherworldly realm at séances, occultism was gaining popularity as an unorthodox and esoteric set of teachings. Occultists were interested in the supernatural for different reasons. More mystically oriented, they hoped to uncover the lost revelations and knowledge of ancient times and make them relevant to the modern world. They claimed that the fusion of contemporary science with sacred and ancient revelations would lead to a new, "complete" science.

Spiritists and occultists were not the only ones interested in supernatural phenomena and mystical experiences. In their own way, turn-of-the-century physicians, psychiatrists, and psychologists were widening their territory of inquiry and developing a sustained interest in the world of the supernatural. In particular, their work on human behavior often led them to consider religious experiences and the manifestations associated with extreme faith. For them, there were no supernatural explanations possible, only physiological and pathological ones. In their eyes, mediums and others were patients, with symptoms that were more like those of mental disease than proof of communication with another realm. More than simple curiosities to classify, mediums and their followers became the proposed basis of theories of mental pathology.

Not everyone agreed. For psychical researchers, mediums exhibited signs of intangible human abilities that could be developed in anyone. In France, psychical research developed around the *Annales des sciences psychiques*, a journal published between 1890 and 1919. The field was based on both observations at séances and the accumulation of testimony from the public. Psychical researchers promoted an open, popular approach to research. Many of them had minimal scientific training and welcomed anyone into their field. By the late 1910s, however, it had become evident that this approach was not bearing fruit, and 1919 saw the creation of the Institut métapsychique international (IMI), a longtime dream of many psychical researchers. From 1919 to 1931, members of the IMI hoped that it would provide the foundations for a serious and respected future science of metapsychics that would explain séances in terms of yet undiscovered powers in humans (telepathy, telekinesis, clairvoyance, etc.). Metapsychists used séances in their attempt to develop a new science of the mind, which they hoped would be incorporated into the scientific corpus.

This story ends in 1931, not because that year marks the end of French interest in the supernatural, but rather because it appears to have been the last time a serious and aggressive attempt was made in France to explain assertedly supernatural phenomena at a more general level. In 1931, members of the IMI failed in their final effort to impress their vision of research on the rest of their community, both nationally and internationally. The 1930s also marked the end of serious consideration of the supernatural by French academics. By then, it had become clear that neither psychical research nor metapsychics would attain legitimacy in universities and other institutions of higher education in the country. For eight decades, however, from the 1850s to the 1930s, the investigation of the supernatural had occupied an uncertain but fertile space of production as unorthodox research holding the promise of potential recognition.

CHAPTER ONE

From Turning Tables to Spiritism

One evening in May 1853, about twenty people gathered at the home of M. Dela-marre, a Parisian banker, to witness an exciting and mysterious new phenome-non. Forming a chain, hands resting on a table, the participants waited patiently for unprovoked movements of the table or sounds of rapping to occur.[1] The Pari-sians had discovered a new game: turning tables. In many salons that season, eve-nings would be spent in a similar manner, waiting for tables to turn of their own accord. The craze, which had already entranced Great Britain and Germany, arrived in France at the end of April that year. Contemporaries talked of a frenzy, an epidemic sweeping Europe. Newspapers and magazines began to feature ac-counts of séances. Pamphlets were quickly put together to explain how to produce the phenomenon to those interested. In just a few weeks, turning tables had man-aged to capture the attention of the whole country. Some authors warned partici-pants against the dangers to a susceptible public of such a futile activity, but for most, table turning was simply an entertaining way to pass an evening.

By June, however, the tables had mostly stopped turning. The press ceased to report on their supposed marvels. The French, it seemed, had lost interest in a game that, if amusing, was limited in its scope. But not everyone abandoned this latest pastime. Over the next few months, as the turning tables continued to fas-cinate a more limited audience, the practice began to mutate. By the end of 1853, what had initially seemed like mere parlor tricks was evolving into a full-fledged pursuit of the supernatural: the séance. The press began to report on this less popular but more complicated trend. At a typical séance, a group of fewer than a dozen people would gather around a wooden table where, connected by their fingers, they would form a chain. Hidden behind two thick curtains, the medium would leave only her or his hands visible on the table. The participants would then patiently wait in the dark, sometimes talking or singing quietly. At a successful séance, the table would at some point begin to oscillate and raise itself to one side before returning to the floor—this signaled that the spirit had arrived and could now be questioned. Reciting the alphabet, participants would wait for the spirit to spell out its revelations through rappings emanating from the table to indicate a particular letter. This method of communication was cumbersome, however, and often the medium would enter a trance, supposedly letting the spirit take control of his or her arm, take pen and paper, and frantically write out a communication. A still more efficient method of communication was for the spirit to enter the medium's body and directly talk to the audience. On occasion, a spirit might also materialize itself, touch some of the participants, and even leave imprints of a hand or a foot in mastic prepared beforehand. Many more phenomena could be witnessed at a séance. Some mediums claimed to predict the future or be able to reveal secrets to their audiences. Others levitated or provoked musical instruments to play, objects to move, or flowers to materialize out of thin air and rain down upon the astonished participants. Whatever their particular specialty, gifted mediums were certain to provide a good show.

Journalists, scientists, and religious thinkers alike became preoccupied with the new phenomena. What lay behind them? Various theories were proposed: some said it was the work of clever con artists; that it was trickery and fraud. Others thought that mediums had mysterious abilities waiting to be uncovered and understood. For others still, divine or demonic forces were at work. From the first mentions of turning tables, a few scientists provided somewhat dismissive explanations for the phenomena they had either heard about or witnessed firsthand. Later on, as parlor games turned into séances and spirits began to manifest themselves and provide what appeared to be tangible proofs of their existence, scientific explanations had to be adapted or changed.

In 1857, using the pseudonym "Allan Kardec," a mathematics teacher from Lyon wrote a book titled *Le livre des esprits*, which introduced his doctrine of spiritism based on messages collected from spirits at séances. In France, Kardec's publication was a defining moment for the phenomena. Thereafter, spiritism became one of the focal points for discussions of séances for believers and skeptics alike. Over the next twelve years, Kardec established a clear set of procedures for both mediums and participants. He built a popular movement through books, a magazine, a spiritist society, and public lectures across France. By the time he died in 1869, his work had won him many followers. Whereas Kardec had emphasized the importance of the messages at séances, the next generation of spiritists would, however, pay more attention to the set of physical phenomena said to be produced by the spirits. Distancing themselves from the religious and moral questions that dominated Kardec's work, they presented themselves as adherents of a rational, even scientific, doctrine. They claimed to possess tangible evidence of the existence of an afterlife. For them, séances combined observation and fact with revelation and faith, promising a reconciliation between the two seemingly opposed poles of science and religion.

EARLY EXPERIENCES WITH THE TABLES

Although attempts to communicate with the dead can be found in many traditions, the modern practice originated in mid-nineteenth-century America. It all began in 1848 in the town of Hydesville in Upper New York State. After the Fox family moved to a new house, the two younger sisters, Margaret and Kate, began to complain of disturbing rapping at night. The Fox parents were soon convinced that the noises were caused by an entity attempting to communicate with their children. Contact with the spirit was established when the adults suggested a method of communication: one rap for "no" and three for "yes." The alleged spirit quickly understood the simple instructions and was able to answer basic questions. Before long, the Foxes developed a more effective method of communication by rapping to the alphabet, which allowed the spirit to relay more substantial messages. He informed the family that he was a previous occupant of the house who had been murdered years earlier and buried in the basement. When bones were duly unearthed, the Foxes and their neighbors became convinced of the authenticity of their spirit. News spread across town and all over New York State. Soon, Kate and Margaret went to live with their older sister, Leah, in Rochester. Not only did the rappings follow them there, but they quickly found they could now communicate with more than one spirit. As the spirits increased in number, Leah Fox began organizing séances in which her younger siblings relayed the messages of the beyond to captivated audiences.[2]

The Fox sisters were the first, but others soon discovered similar abilities to communicate with the dead. A new movement developed around them: spiritualism. Spiritualists claimed to have found definitive proof of the afterlife through these communications with the other world. They believed that séances made it possible to base religious beliefs on physical facts rather than faith alone. Concrete evidence was at the heart of spiritualism. As the providers of observable and verifi-able links between this world and the next, séances and mediums came to occupy a pivotal place in the spiritualist practice.[3]

As the American spiritualist movement grew in importance, news of its activi-ties spread to Europe. In the summer of 1852, the French Catholic newspaper *L'univers* reported on the notable progress of this new sect.[4] But it was traveling mediums who would introduce Europeans to the practices associated with Ameri-can spiritualism. The first European séances took place in Scotland. From there, the craze is said to have taken hold of the continent, city after city.[5] Reporting from Bremen in March 1853, a German physician described the atmosphere following the arrival of a steamer from New York: "For about eight days now, our good city has been in a state of agitation difficult to describe. . . . There is not a house around here in which people are not busy with this fantastic exercise."[6] By April 1853, the inhabitants of Strasbourg, Marseille, Bordeaux, Toulouse, and every other major center in France were learning how to channel the spirits and make tables turn. Everyone it seems was fascinated by this new game.[7]

How-to guides were rapidly put together to provide an interested public with information on the subject. The writer and journalist Ferdinand Silas offered a series of detailed recommendations on how to reproduce the phenomenon suc-cessfully. For good results, he insisted, there should be no more than five partici-pants, men between the ages of eighteen and twenty and women between the ages of sixteen and forty. Family members and friends should sit next to one an-other. A light, oval wooden table with rollers should be used and placed on carpet so as to facilitate its movements. Participants should form a chain, connected by their fingers, and avoid constricting the movements of the table. If sufficiently patient and focused, they should begin to experience unfamiliar sensations, such as heat and tingling in the fingers, arms, and chest. After twenty to sixty minutes, they should see the tabletop oscillating. Then, if the human chain around it re-mained closed, the table would slowly begin to rotate in the direction previously agreed upon.[8]

If the table-turning mania seems to have taken hold of the French almost overnight, it was abandoned just as quickly. Limited to a repetition of the same procedures with the same outcomes, turning tables soon lost its excitement and interest. After all, how many times did it take before the novelty of seeing a table turn by itself wore off? On June 25, after only two months, the weekly newspaper *L'illustration* declared the tables to have fallen still.[9] The fad had ended. "Every-thing exhausts itself in this world," the popular science writer Louis Figuier later recalled. "When it had been repeated a sufficient number of times, we grew tired

Early séances were limited to table-turning in social settings. From Louis Figuier, *Les mystères de la science* (Paris: Librairie illustrée, 1860), 1: 73.

of this occupation, in fact, fairly dull, which added nothing to what we already knew. The tables thus stopped everywhere and with one accord."[10] Gathering around turning tables did not cease entirely, but the practice shifted to a more complex set of phenomena, reminiscent of the American spiritualist séances. Talking tables were harder to find. They required more than just enthusiastic participants eager to obey instructions: a medium who could summon spirits and communicate with them was needed.

There is little doubt that many people would have found these first conversations with the talking tables tedious. Participants laboriously recited the alphabet, waiting for a rap to indicate a specific letter and slowly form words and sentences. Once suggested, the technique of automatic writing allowed to speed things up. A medium, possessed by a spirit, would write or dictate messages from the beyond. With newer and faster methods of spirit communication, exchanges became longer and more complex. Séances were becoming interesting. They began to take place in numerous homes, even those of the rich and famous. At court, Empress Eugénie organized séances and shows involving mediums. While in exile on the Isle of Guernsey, Victor Hugo also experimented with the tables. With his family and his friend and fellow writer Auguste Vacquerie, Hugo supposedly communicated with the spirits of Racine, Molière, Dante, and others. Exchanges with past literary figures, however, failed to provide much satisfaction for the author, who soon came to the disappointing conclusion that poems and other written works that had been obtained during séances were more likely to be the unconscious creations of the participants rather than those of deceased geniuses. If something did hold Hugo's interest in séances during his time in exile, it was the communications with his beloved deceased daughter Léopoldine, who had drowned a few years before. In his interactions with her supposed spirit, the author is said to have found some solace.[11]

Hugo's interactions with the spirits of famous men were the norm rather than the exception. Across the country, many historical figures appeared at séances and were given the opportunity to tell their side of historical events. Recalling séances taking place in 1853, for example, Henri Carion, chief editor at the ultra-conservatist *L'émancipateur de Flandre et d'Artois*, remembered a conversation he once had with Jeanne d'Arc in which he had asked about the fate of Louis XVI and his family. They had safely made their way to heaven, she had reassured him, including the dauphin, who had not survived imprisonment.[12] Jeanne d'Arc was a particularly popular spirit in those early years. So were Napoléon, François-René Châteaubriand, Jean-Jacques Rousseau, and Saint Augustine. Each spirit brought particular messages to séances, participating in historical, political, social, moral, religious, and even scientific debates of the period. Through mediums, the spirits could divulge any piece of knowledge or surprising information to participants.[13]

As such, séances held great potential for mediums. If spirits possessed the power to rewrite the past and predict the future, if they could alter political views or shape the knowledge of their audiences, mediums necessarily played a part in this. They were, at the very least, essential to séances and, at most, the creators of this new knowledge, whether consciously or not. Mediumship had its benefits. In France, as earlier in America and Britain, many participants, particularly women, rapidly discovered their own mediumistic abilities. From those who performed in small, private circles to those who traveled from city to city, accepting invitations to the homes of the wealthy, mediums' talent earned them respect and sometimes money.[14]

Mediums were not the first to claim supernatural powers, even in their own century. The tables began to talk in a world already full of extraordinary manifestations and mystical women and children. Within the Roman Catholic tradition, stories of visionaries and Marian apparitions fascinated both the public and the press. In 1830, claims of Mary's apparitions to a nun in Paris led to the sale of millions of medallions around the world. In 1846, in the small town of La Salette in the French Alps, two young shepherds, Mélanie Calvat and Maximin Giraud, described visions of a weeping Mary, a claim that rapidly attracted pilgrims and was eventually promoted by the Church. A decade later, in 1858, Mary appeared seventeen times to the young Bernadette Soubirous, transforming Lourdes into an important site of pilgrimage and miraculous healing, which it remains to this day. Outside the religious sphere, strange phenomena were sometimes said to be taking place around adolescent women. A few years before the French began talking to the dead, for example, the teenager Angélique Cottin was mentioned in the press on numerous occasions in relation to the claims that her presence caused furniture to fly. Just like mediums would after her, the "electric girl," as she was dubbed, attracted a lot of attention but few explanations. In 1846, she traveled from her native Normandy to Paris, where she met with eminent scientists of the day and became the subject of a discussion at the Académie des sciences. A commission headed by François Arago was unable to confirm the phenomenon, however, and failed to submit a report.[15]

Closer to mediums in their abilities were somnambulists, who by midcentury had already gained notoriety for their clairvoyance and other talents. Somnambulism was associated with animal magnetism, the belief that a fluid residing in all living beings could be manipulated in particularly susceptible subjects. Somnambulists were mostly women—with some notable exceptions—working with their magnetizers, usually men, who were portrayed as strong, full of vital fluid and able to command others. In public performances, they entered a trance, during which they were said to show impressive abilities such as thought transference and sometimes even communication with the dead. Some somnambulists received clients in their homes and earned a living through their consultations. As late as the beginning of the twentieth century, somnambulists were even mentioned in guidebooks to the city. The 1906 *Guide des plaisirs à Paris*, for example, lists the addresses of nine female somnambulists recommended for second-sight consultations, along with other clairvoyants.[16] By then, however, although some somnambulists still continued to practice that trade, the large majority of them had made the transition to mediumship. In a setting in which séances dominated other types of supernatural production, those with talents of the unexplained and mystical kind generally followed the trend and adapted.[17]

FEARS AND EXPLANATIONS AT THE SÉANCE

From the beginning, séances led to many pamphlets and commentaries. The fact that they took place in an atmosphere of excitement and sometimes even sensual tension did not go unnoticed. Participants waited for a supernatural manifestation in a dark room. Men and women sat in alternate order around a table, hands brushing each other. In a poem, the army captain Jacquet remarked on this in telling the story of a respectable countryman who, on a visit to Paris, had witnessed a séance in a salon. There were great dangers in such a practice, the countryman claimed. The touch, the excitement, and the mystery could bring confusing feelings of love. Fathers should be warned and prevent their daughters from becoming involved in such a promiscuous situation or suffer the consequences.[18]

As séances developed from simple parlor tricks into sites of communication with the other world, mediums became the focus of the practice. Most scientists dismissed their performances as either conscious or unconscious fraud. Critics of séances often mentioned the dangers to the mental health of participants, citing cases of madness developing in some of them.[19] The journalist Ferdinand Silas warned of the side effects that could occur—fatigue, malaise, and nervous excitement were often experienced during séances. In most cases, these sensations stopped shortly afterward, but in the most sensitive participants, they could become permanent.[20] The pharmacist P.-F. Mathieu was also concerned for the well-being of participants and mediums. The tables often confused, and this sometimes led to madness in the more sensitive attendants and mediums, Mathieu wrote. He advised everyone involved, particularly the more emotionally fragile, to be cautious.[21] Physicians also reported physical hazards involved in table turning and séances. A medical journal from Strasbourg even reported accidents and, in one case, a death, due to flying furniture.[22]

The Catholic Church had different fears and with good reason. This was not the first nor would it be the last time that it was confronted with competing claims of the supernatural. Séances, however, were more than just another supernatural manifestation. They offered a potentially very powerful rival authority on both earthly and spiritual matters. Previous challenges such as cases of visions or possession had always remained discrete events, limited in number and short in duration, but séances were reproducible. They could take place anywhere and at any time. They provided open-ended teachings, since spirits might always reveal new truths to their audience. Moreover, the revelations were produced in ways that were presumably observable and testable. As such, mediumistic phenomena held the potential to make faith obsolete and provide proofs stronger than religious revelations.

A few Catholic authors believed the phenomena to be authentic and suggested that Satan was behind them. Some wrote that séances were providing the devil with a new weapon in his war against God on Earth. For one anonymous cleric, the fact that the tables became agitated, dropped, and refused to answer questions about Jesus—even when interrogated—was a strong indication of Satan's involvement.[23] For the most outspoken writer on the presence of Satanism at séances at the time, Jules-Eudes Mirville, the tables were a symbol of materialism, emblematic of the arrogance of an era that failed to accept the explanatory power of spiritual causes. By the time the tables began to move in France, Mirville had already established a firm reputation as a crusader against what he felt were the various supernatural manifestations of the devil, focusing particularly on claims of haunting. For him, only the activity of an intelligent agent could explain the phenomena, and that agent had to be the devil. As for the tables, they were, at the very least, the expression of a soulless society and, at the worst, the work of Satan.[24]

Other Catholic writers were more hesitant in attributing a cause to the talking tables. The philosopher André Pezzani did not believe that Satan was directly involved in the movement of the tables, but thought their motion was caused by the voices of spirits acting without direction or purpose. Pezzani feared, however, that the confusion produced would shatter Catholic unity and thus contribute to the devil's mission on Earth.[25] For Pezzani, Christians who heeded the tables were exposing themselves to dangerous ideas because the tables revealed that there was no difference between humans and animals, that life had supposedly begun in the plant kingdom, and that organisms had evolved from lower life forms to increasingly complex ones and finally to civilized beings.[26]

Most Catholic authors agreed with Pezzani in rejecting diabolical explanations for the manifestations, even if they did believe in the authenticity of the phenomena. A séance could be a wonderful experience, allowing a believer to reconnect with a deceased friend or family member. Spirits could be angels or dead souls from Paradise, Purgatory, or Hell; or they could be the souls of children who had died before baptism. For the Catholic journalist and journal editor Henri Carion, the nature of the spirit had to be established while conversing with the entity. Celestial spirits would be wise and concise, mischievous spirits would show pride and a confusing discourse; good spirits would give sound advice and be charitable; souls in Purgatory would beg and humbly confess their misery, and the damned would be recognized by traces of the violent passions that had led them to where they now were. Good spirits, Carion stated, refused to answer frivolous questions.[27] For P.-F. Mathieu, it was the responsibility of those at a séance to establish a spirit's identity. Spirits could be a source of consolation and knowledge, but not all of them should be trusted. Whether Satan was speaking or not, séances should be conducted with seriousness and extreme caution for fear of physical, emotional, even spiritual injury.[28] The Parisian abbé Almignana considered the possibility that Satan was behind the moving and talking tables but rejected the hypothesis after having failed to observe signs of possession. Believing in the authenticity of the phenomena nonetheless, he concluded that the voices heard had to be those of the dead.[29]

Catholic authors were not the only ones who wrote on séances. From the very beginning, scientists were asked to observe and comment on the flying furniture and the moving and talking tables. In late May 1853, a séance participant, one de La Giroudière, published a letter in which he called for scientific discussion of the phenomena. He deplored how in the past scientists had refused to consider certain more puzzling manifestations. Why, he asked, did they not study the marvelous in the same way that they studied new elements in chemistry? Why did they not observe a new phenomenon, experiment with it, and record its actions and influences before finally classifying it and assigning it a place within science?[30] Ernest Bersot, director of the École normale supérieure, reminded scientists that their discipline had its roots in the marvelous, a claim often made in these sorts of arguments. Science was born of inexplicable facts; from a combination of the natural and the marvelous, it had slowly developed into a set of scientific laws, he wrote. If moving and talking tables were deemed supernatural in nature, it was only temporary. After rigorous observation and scientific explanation, they would also be recognized as natural phenomena.[31]

The majority of scientists refused to answer the call for a serious investigation. Only a few showed any interest in the mysterious tables throughout the 1850s, both in France and abroad. The earliest and most prominent of them was the physicist Michael Faraday, renowned for his contributions to electricity and magnetism. He gave a mechanistic explanation for the movements of the tables. Small movements of the hands, provoked by the will of the participants, made the tables rotate. Faraday's comments on the moving tables were published in the reputable British magazine *The Athenaeum*, and in July 1853, a translation of his article appeared in the French weekly *L'illustration*. It presented the first version of the explanation that would be most often invoked to explain séances in their early days.[32]

As for the Académie des sciences, it was required to produce its own study on the topic after it received reports on the divining rod and the turning tables soon after their appearance in France. A commission was created and headed by Michel-Eugène Chevreul, a chemist at the Muséum d'histoire naturelle. It was not the first time Chevreul had been asked to investigate peculiar occurrences. In 1833, he had been on another commission of the Académie to consider the strange movements produced by pendulums when held by the string. At the time, he had provided a mechanistic explanation similar to that of Faraday years later: small, unconscious muscular movements by the subject could produce visible and mystifying movement in an object.[33] It was the same unconscious muscular movements, Chevreul explained, that produced the phenomena associated with the divining rod and the turning tables.[34]

If Chevreul did not see enough in this to attract his interest beyond the report of the commission, séances did manage to attract the sustained interest of another member of the commission. In 1854, the physicist Jacques Babinet wrote two articles for the *Revue des deux mondes* on the tables. Like Chevreul, Babinet was no stranger to manifestations of the supernatural. In 1846, he had been part of the commission on Angélique Cottin, the electric girl; and, in 1851, he had taken an interest in claims of a haunted house in the Norman town of Cideville. Science was based on facts, he explained. Were there undisputable facts in these cases? Previously, the lack of scientific evidence had made him reject the existence of phenomena such as the electric girl or the Cideville haunting. But in 1854, when Babinet considered the tables, he found himself believing that they could be genuine, explaining their movements as the product of a combination of simultaneous nervous trepidation on the part of the many participants around them.[35]

By then, mechanistic explanations of the phenomena had become increasingly popular. Darius Rossi, a self-professed "professor of literature," developed a law of "parallelogram forces" to describe the movement of the tables. The sum of the unconscious movements of each participant and the tendency never to counteract the initial movement resulted, for Rossi, in a parallel force that produced a change in the table. In a small pamphlet, Armand Maizière, a prolific writer and corresponding member of the Institut historique de France and the Société météorologique de France, took another approach, arguing that the movement of the tables resulted from heat. At each point of contact between the table and a participant, the table accumulated caloric energy. Once the accumulated energy became greater than the friction between the participants' hands and the table, a motion would begin and be pursued by the determined participants. A study of the caloric factor, Maizière concluded, was the key to understanding the phenomenon.[36]

The talking tables were, of course, more difficult to explain. Babinet believed them to be simple deceptions—a ventriloquist's trick. Chevreul suggested that they were the expression of the unconscious thoughts of mediums. An anonymous author writing under the pseudonym Grosjean came to a similar conclusion. Grosjean believed that the motion of the table rose out of the unconscious, and that the thoughts expressed were those of the participants. Thus, what might seem like new ideas were either forgotten memories or a new combination of previous ideas. Grosjean argued that every person harbored two personalities. Gifted mediums were not impostors or charlatans but individuals who could separate these two personalities for a short period of time using a potential that lay dormant in everybody else. While in their second personality, they had the ability to read the thoughts of others, even at a great distance. Grosjean's theory of separate personalities did not attract a lot of attention at the time, but the psychologist Pierre Janet would later see in it an embryonic version of his own theory of the dissolution of personality.[37]

Many thought that the phenomena, if real, were the product of the medium's mind, but was it a healthy or a sick mind? For Alphonse Chevillard, professor at the École des beaux-arts, mediumship was a self-induced nervous disease that everyone held the potential to develop.[38] Nervous emanations originating in the unconscious of the diseased individual were responsible for the phenomena. As such, séances were a reflection of some unexplored regions of our physiology. They called for a new theory of mechanical magnetism in which the phenomena could be reduced to the unconscious manifestations of a magneto-dynamical and nervous fluid.[39] Thus, by the late 1850s, a number of explanations of séances had already been formulated in France. Whereas the moving tables could easily be accounted for in terms of small movements, whether provoked by the unconscious will of the participants or by a magnetic fluid, the talking tables could not be so easily dismissed. Hypotheses involving fraud, separation of personalities, and transfer of fluids were suggested, but none of these explanations seemed to gain wide acceptance. In 1857, however, a new explanation for the talking tables would be proposed: spiritism. From then on, it would provide a focus for the discussions of both supporters and opponents of séances in France.

KARDEC AND SPIRITISM

Once the tables began to talk, it seems that they had a lot to say. In the first few years, the wealth of knowledge that accumulated as a result of conversations with supposed spirits lay in a disorganized state. There was no dominant framework from which to interpret the numerous messages received. The meanings ascribed to séances varied from group to group and among believers. One of the earliest attempts at a doctrine based on the messages of the spirits was compiled by Louis-Alphonse Cahagnet, an enthusiastic Swedenborgian cabinetmaker. By 1848, Cahagnet had already published the first of three volumes on his work with somnambulists and their conversations with the dead. In 1854, he turned his attention to séances and encouraged participants to send to him transcripts of their attempts to communicate with the spirits. In his participatory approach to séances, Cahagnet invited anyone to contribute to his journal, the *Encyclopédie magnétique spiritualiste*. Any explanation from any walk of life or mode of communication would be accepted and published. It was up to readers to come to their own conclusions about the messages. The *Encyclopédie magnétique spiritualiste* was not successful. Bitterly explaining its failure, Cahagnet lamented most people's unwillingness to hold personal opinions. In his own words, the inevitable failure of his journal had been due to his disordered approach. Man was weak and needed guides, he concluded sadly.[40] Where Cahagnet had failed, others tried. Baron de Guldenstubbé, an enthusiastic Norwegian séance organizer, made his own attempt at understanding the messages a few years later. In the talking tables, he saw a way to obtain religious truths. The writings produced at séances would provide the scientific basis of the phenomena and their moral consequences. Like Cahaget's before him, Guldenstubbé's views obtained little success.[41] In the end, it was Allan Kardec's 1857 *Le livre des esprits* that would provide the most widely accepted explanation of séances in France.

Before he became Allan Kardec, the future father of spiritism was Denizard-Hippolyte-Léon Rivail, a mathematics teacher from Lyon. Born in 1804, Rivail spent his early years in France before relocating to Switzerland with his family. There, he attended the school of the Swiss educational reformer Johann Heinrich Pestalozzi in Yverdun. This experience would later shape his own conceptions of education and influence his moral ideals. As an adult, Rivail moved to Paris and founded an Institut technique, where he adopted the educational methods of his earlier teacher. In 1832, he married a fellow teacher, Amélie Boudet. For a time, he seemed to enjoy a comfortable and happy life, but soon financial problems

Allan Kardec, the founder of spiritism. From Sir Arthur Conan Doyle, *The History of Spiritualism* (London: Cassell, 1926), 2: 172.

arose. His uncle, who had provided funds for the school, gambled away his money, forcing his nephew to close his institute. Rivail continued to feed his passion for education, however, by publishing numerous pedagogical treatises.[42]

During those years, the young educator became fascinated by all phenomena related to magnetism. Later on, he would remark that this background had prepared him well for the study of spiritism. In fact, it was through his interest in magnetism that Rivail discovered séances in 1854. After repeatedly hearing about the feats of the tables from many friends, he is said to have reluctantly agreed to attend a séance in 1855. The experience must have been convincing, because he soon joined a group and began to attend on a weekly basis.[43] From there, everything progressed swiftly. Asked to put some order into a series of communications obtained within his group, Rivail began his work on the messages of the spirits. He quickly decided that the spirits summoned had no particular authority or wisdom. One had to take their teachings as opinions. Under his influence, séances held in his group began to change. Questions became more direct and specific. Rivail wanted to produce a guide for practitioners that would help them, not only in their séances, but in their beliefs, their scientific outlook, and their daily lives. In this task, he was guided by a particular spirit, whom he referred to as *la Vérité* (the Truth.)[44]

The *Livre des esprits* came out in 1857 and introduced spiritism to the followers of séances. Spiritism, for Rivail, was more than a religious doctrine; it was a philosophy, spiritual in its concerns but grounded in hard observable scientific and rational principles. Rivail had worked with a dozen mediums in order to assemble the information presented in his book. He published it under the pseudonym of Allan Kardec — a name that he claimed had been revealed to him as having been his own in ancient times, when he had been a druid. From then on, Rivail became Kardec, founder of the doctrine of spiritism. The *Livre des esprits* was a hit. The first edition sold out quickly. In 1858, a second, more substantial one was published. Further editions came out in 1860 and 1861. By 1874, the book had already gone through twenty-two editions. Today, it is still the standard reference text among practicing spiritists as it continues to be republished in French as well as in other languages including English.[45]

For Kardec, spiritism was neither new to his century nor the discovery of two young American girls. On the contrary, moving and talking tables dated back to ancient times. The spirits had been conversing with the living for centuries. The Fox sisters had simply rediscovered what had already been known and accepted by the great minds of the past. He claimed that, among others, Pythagoras and his followers had been precursors to the doctrine. Something had changed, however. The modern séances had developed in a new era of which reason and progress were dominant trends. This had led spirits to reveal new teachings more in keeping with the times, such as the principle of progressive reincarnations. Ancient Indians and Egyptians had believed in metempsychoses, that souls could be reincarnated in human or animal bodies. Spirits revealed to Kardec, however, that transmigration from a human to an animal body was impossible. Reincarnation always happened in a progressive direction.[46] As such, spiritism embodied progress, the ultimate goal for humans, realized in a series of reincarnations, each one allowing for the expiation of past faults and leading humans toward improvement. For Kardec, this principle of progressive reincarnations made the spiritist doctrine moral because it gave a sense of justice and purpose to the lives of believers.[47]

In the *Livre des esprits*, Kardec described spiritism neither as a science nor as a religion, but as a philosophy. His methods of research and arguments were pedagogical and his presentation reminiscent of a system of geometry. The book opened with an introduction to the terminology and a few basic concepts (spiritism, God, infinity, soul, spiritist, etc.). Kardec then presented his doctrine in the form of questions and answers, and from there slowly built his doctrine by asking increasingly complex questions. The format gave the reader the impression that the spiritual truths presented were comparable to theorems and mathematical truths. At the same time, spiritism by nature was open-ended knowledge. The *Livre des esprits* was never meant to be the final word on the subject. On the contrary, spiritism was a set of teachings that could grow with and through séances. Future conversations with spirits could lead to new knowledge. Any group holding séances could potentially contribute to it, and Kardec invited his followers to participate by sending in records of their own séances. Of course, this put his doctrine in a precarious position. The spirits were unreliable and often provided contradictory messages. For spiritism to be a coherent doctrine, an efficient way of screening the various messages had to be established.

In April 1858, Kardec founded the Société parisienne des études spirites (SPES), with branches in several French cities. Every Tuesday night, the SPES held a séance in the Galerie de Valois at the Palais-Royal in Paris with the medium Ermance Dufaux. The room could contain around twenty individuals. To prevent unexpected revelations, Kardec provided clear and established procedures for the reception and dissemination of messages: silence was mandatory; all questions had to be approved by the president before being put to a spirit; and futile, personal, or tricky questions were forbidden. Spirit communications received outside the society would be read only after they had been submitted to the president or committee for approval. Occasionally, a guest could be admitted, but only if introduced to the president ahead of time and recommended by a member. The SPES became so popular that a more spacious locale soon had to be found, and the society thus began to meet on Fridays at the Dious restaurant in the Galerie de Montpensier at the Palais-Royal. After 1860, it would meet in its new headquarters in the passage Sainte-Anne close by.[48]

To centralize and strengthen his doctrine and exercise further control on the incorporation of new revelations into its accepted corpus, Kardec launched the *Revue spirite* in 1858. The journal relayed stories from the growing movement, news of séances around the country, and new accepted teachings from the spirits. Initially, the *Revue spirite* consisted for the most part of moralizing essays on various topics and of conversations with famous deceased personalities. Jeanne d'Arc and past kings of France featured prominently in its pages. So did past artists and geniuses—Wolfgang Amadeus Mozart, for example. Kardec kept a tight control over the content of his journal. Every essay or transcript of a séance had to be approved by him before publication. Throughout the 1860s, the *Revue spirite* had as many as 1,800 subscriptions.[49] In 1860, Kardec also began to travel across France to promote his doctrine. The lectures he held never drew significant crowds, but they did solidify the movement across the country.[50]

Beyond the book, the society, the journal, and the lecture tours, spiritism remained first and foremost a practice, and the practice needed methods, rules, or steps that could be codified. In 1861, Kardec developed the procedures to be followed at séances in greater details in *Le livre des médiums*. Whereas the *Livre des esprits* had focused on the doctrine in its moral and philosophical implications, the *Livre des médiums* gave a how-to guide for séances. Kardec documented each phenomenon one could expect to witness at a séance and the manner in which it could be obtained. Although he listed many physical phenomena, the priority was given to conversations with spirits. In the *Livre des esprits*, Kardec had differentiated between an experimental spiritism, dealing with spirit manifestations in general, and a philosophical spiritism, dealing only with intelligent manifestations. The spiritist doctrine was a product of the latter, not the former. For him, séances were serious gatherings, where communication with the beyond was established, preferably through automatic writing. They were sites of knowledge production and learning, not of entertainment and spectacular witnessing. In fact, in the rules of the SPES, there was no mention of physical phenomena, only of conversations with spirits obtained through a medium and accepted by a president elected every three years.[51]

For Kardec, spiritism provided the only valid explanation for what occurred at séances. As long as spirits had limited their manifestations to physical phenomena, Kardec argued, the tables had remained unproblematic for science. Motion could easily be explained mechanistically and did not require an appeal to the occult. Fluids such as electricity could also convincingly account for moving tables and even lead to interesting scientific theories. When the tables had begun to talk, however, physics and physiology had become insufficient to account for the phenomena. Science dealt with the observation of material facts; as soon as intelligent causes were involved, it was at a loss.[52] Outside the realm of science, spiritism provided a new and unexplored method to study intelligent causes. It also constituted the foundation on which a new science would rise, one based on both reason and the revelations of the spirits.

On scientific matters, the spirits dispensed teachings on everything from the smallest particles of matter to the infinite universe. Their privileged access to knowledge could guide science and inform scientists. They taught Kardec that matter is made up of a sole element and that everything is a modification of this elementary molecule. They revealed that everything could be reduced to force and the motion of matter. They told him about the origins of the universe, the Earth, and living beings. God's will, forces and vital principles, and fluids organized life. Spirits provided answers to the day's debates: for example, spontaneous generation was possible, the Earth was still very young, races were a product of different climates and habits, and other worlds were inhabited. They confirmed that Adam had indeed existed, but that he had neither been alone nor the first human. They revealed that all of humanity shared in the same goals of improvement and progress. Equality in goals, however, did not mean true equality. Spirits taught believers that the souls of indigenous populations were still at an early stage of development, but deserved the respect of so-called civilized people, who were after all still in many ways savages themselves. Spiritism was thus a doctrine in which progress was a phenomenon both of matter and spirit. Humans were made up of elementary molecules and vital principles. They evolved through reincarnation from primal to enlightened beings. As children developed into adulthood, their souls progressed slowly from a burgeoning intelligence in a creature of passions to a civilized being, finally reaching the stage of an advanced spiritual being.[53] In their focus on progress and reason, spirits were entrenched in their time. Through their discussion of many scientific issues of the moment, they demonstrated a desire to participate in the scientific culture of the world surrounding them. Their relationship to science, however, remained ambiguous.

Over the years, Kardec increasingly presented his doctrine as scientific. By the time of the 1861 publication of the *Livre des médiums*, spiritism was no longer a doctrine or a philosophy guiding science; it was both a philosophy and a science. What Kardec meant by this, however, was unclear. The scientific method, for example, was still said to remain inapplicable to spiritism. In fact, if anything, his study of spiritism worked in the opposite direction.[54] Spiritism was scientific, but it was not an observable or an experimental science. It derived from reason in the first place and only then, perhaps, from the senses. Such was the true path to knowledge, Kardec wrote. As a former mathematics teacher, Kardec valued rationality. His books were intended to appeal to reason. He presented his doctrine in a manner he judged to be logical. He obviously believed in the promises of rational knowledge, and this was a large part of what attracted him to séances: spirituality based on concrete manifestations. If he hoped to obtain a scientific status for the manifestations at séances, he never meant them to be dependent on it. He saw them as true and real outside of science, demonstrated by rational principles and observations. It was science that would be modified by them and not the other way around. When he wrote that spiritism was both science and philosophy, he most likely meant that it was a fact in the scientific sense, but that it existed above science, and would eventually modify it through its inclusion. Kardec meant that spiritist knowledge in its most profound way could influence scientific knowledge, but not be influenced by it, that it was the highest form of knowledge available.

SCIENTIFIC SPIRITISM

Kardec limited membership to the SPES to believers only. This meant that, during the 1850s and 1860s, it was almost impossible for interested but skeptical observers to witness spiritist séances in France. If there were no unconverted scientists admitted to the proceedings, however, a few members of the society were evident science enthusiasts. The most famous of these early scientifically inclined spiritists was Camille Flammarion. Never formally trained as a scientist, Flammarion would go on to have an incredibly successful career as a popularizer of science. In the early 1860s, he was earning a living as an assistant at the Paris Observatory and devoted most of his free time to his scientific writing. In 1862, his passion for astronomy and his belief in extraterrestrial life led him to write a book titled *La pluralité des mondes habités* (The Plurality of Inhabited Worlds), in which he described life as the ultimate purpose in the universe. That life flourished on Earth was visible everywhere; the same would ultimately be discovered about the rest of the universe, he thought.

It was while doing research for *La pluralité des mondes habités* that Flammarion came across a copy of Kardec's *Livre des esprits* in a bookstore. Noticing the discourse of the spirits on the plurality of worlds, he became intrigued by the doctrine. According to Kardec, all planets held life, all worlds were united, and reincarnation was a universal phenomena. As conditions were different for each world, inhabitants varied from one planet to another, but behavior in one existence determined circumstances for the next one whether on the same planet or on a new one.[55] For Flammarion, the *Livre des esprits* was an interesting discovery to be included in a section of his book on the various beliefs associated with the plurality of worlds over the centuries.[56]

There was another reason for Flammarion's interest. Spiritism was a doctrine grounded in observable facts, he thought. Believers claimed that the immortality of souls was proven at each séance. Trusting in both the centrality of life in the universe and the great promises of science, Flammarion took a serious interest in the movement. He was convinced that life was eternal and could be found everywhere, but he refused to accept this on faith alone. Although brought up Catholic, he had, over the years, become increasingly dissatisfied with the Christian faith. In his memoirs, he later explained: "Scientific history is founded on direct observation of facts in nature, whereas religious history offers at its base pure fictions only, naïve, indemonstrable, and even contradictory."[57] By the 1860s, Flammarion had turned toward a more naturalistic spirituality and had come to believe that proof of God's existence was apparent in the nature around him, that science was the only possible path to this truth and to God.[58]

It was Flammarion's contention that if God could be felt in all of nature, he would be especially noticeable in the heavens. As such, astronomy was particularly important, and its unification with spiritual concerns would remain one of Flammarion's constant preoccupations: "It is indispensable that the system of a moral world and the system of a physical world come to form a single unity; astronomy and religious philosophy must be in harmony, and I believed myself forced by the direction of my studies themselves to establish and to demonstrate this truth."[59] In spiritism, he saw the promise of this unification: a faith in immortality and the plurality of worlds based on concrete and observable facts.[60]

Flammarion spent the early part of 1861 familiarizing himself with the doctrine. In June of that year, he wrote to his friend Charles Burdy: "I have just reread Allan Kardec's book. I have found in it a profundity of thought and I think him of good faith. Thus it must be that either spiritism is not a utopia [i.e., is not incredible] or this thinker is crazy."[61] By October, Flammarion had met Kardec. In a letter to the abbé Collin, he related his first encounter with the spiritist and his intentions regarding the movement:

> I went to see the chief of the spiritist school, the spiritist par excellence Allan Kardec; I have done everything to implore him to receive me in his private séances, where mediums (apparently) communicate with the spirits; the said Allan Kardec affably received me and appeared to me, contrary to all expectations, quite *foreign to the spirit of the system*. I did not disguise the motives that brought me to him, making it clear that I only want to be convinced by my reason, that I have no preconceptions . . . in this respect, because I cannot yet tell where the truth is to be found, that my only desire is to have before my eyes proofs that if they are valid will make me a follower and a propagator of this doctrine, and if they are worthless will render me the eternal enemy of a utopia that can serve only to mislead weak minds.[62]

In a subsequent letter to Burdy, Flammarion described his first séance at the SPES and the uncertainty with which he was left: "What do you think of this? As for myself, I am always on the trail of tricks and strings: I would only want to believe at the last moment possible, that is to say in the strongest way possible."[63]

Over the next few months, Flammarion remained cautious, but his interest in spiritism continued to grow. Although he would never become a true, convinced spiritist, his ties to the movement rapidly strengthened. On November 12, 1861, he wrote to Kardec asking to be formally accepted as a member of the SPES. Flammarion described his own interest in spiritism as follows:

> Impressed with a profound respect for the teaching of spirits and filled with a just admiration for the immortal fruits that this new branch of psychology has already borne, my greatest happiness would be to be able to delve into this doctrine as much as possible. Christian studies having taught me the sacred dogma of the immortality of the soul, and astronomical studies having made me a fervent believer in the plurality of worlds, I was led by these beliefs to the doctrine of spiritism, of which they form the double basis, and I would be happy to be able to continue the studies I have begun on this science fruitfully.[64]

Three days later, Flammarion was officially invited to join the society. In his request to Kardec to become a member, his position on spiritism remained clear: spiritism lay between science and religion, holding out the promise of a unification of faith and reason. A year later, Flammarion wrote *Les habitants de l'autre monde*, a book consisting mostly of mediumistic writings, in which he enthusiastically declared the phenomena to be real.[65] This first public endorsement marked the beginning of a lifelong interest by the astronomer in séances and their manifestations.

Although Flammarion would continue to study spiritist phenomena until his death, he did not feel as strongly about the spiritist movement itself. Over time, his ties with Kardec's society weakened. By the mid 1860s, he was developing an interest in the kind of physical productions that were discouraged by Kardec and moving away from the spiritist explanation. He began to develop the concept of unknown natural forces to account for some of the phenomena observed at séances. His first mention of naturalistic causes dates from his account of the visit of two American mediums, the Davenport brothers, to Paris in 1865. At the time, the brothers were famous for an act in which they were tied up inside a cabinet and appeared to be playing musical instruments. If their performance had been greatly appreciated by the public—they had even been invited to perform for the emperor's family—they had been received coldly by the press. The brothers' mediumistic faculties had even been called into question by the *Revue spirite*, which, at the time, did not encourage physical manifestations.[66] Under the pseudonym of Hermès, Flammarion had written an account of the Davenports' visit. Although he had not defended them as mediums, he had believed in the authenticity of their act and had preferred to write of unknown natural forces rather than spirit intervention to explain their particular abilities.[67]

Kardec died in 1869 leaving no clear instructions regarding a possible future leader for the movement. There were a few candidates, including Flammarion. At Kardec's funeral, however, the popular astronomer's eulogy left little doubt about his wish to distance himself from the movement. Asked to preside over the ceremonies by Kardec's widow, he reluctantly spoke of his views on the future of spiritism. He had now come to believe that the spiritist doctrine lacked rigorous proof. It was time for spiritism to enter its scientific phase, he said, which could only be accomplished through the study of the diverse physical phenomena experienced at a séance. Manifestations of the spirit world needed to be dissected, measured, and defined. Spiritism was not a religion, not a dogma, he declared; it was a science of which we knew nothing yet.[68] "The supernatural does not exist. . . . There are no more miracles. We shall witness the dawn of an unknown science. Who can predict to what consequences the positive study of this new psychology will lead," he almost sacrilegiously declared on the tomb of the leader.[69]

In the end, Pierre-Gaëtan Leymarie and his wife Marina replaced Kardec as editors of the *Revue spirite* and founded the new Société anonyme pour la continuation des oeuvres spirites d'Allan Kardec.[70] Leymarie's vision of spiritism was closer to that of Kardec, but he shared Flammarion's conviction that the future of spiritism lay in science. Under his direction, the *Revue spirite* reproduced significantly fewer reports of communications with spirits and allowed more space for physical phenomena and news of the movement. Leymarie also pushed for photography as a method of control and research at séances. As moving objects and materializations did not run the risk of contradicting the established spiritist corpus in a way that written communications with spirits could, the spiritist movement grew less centralized and easier to penetrate. Spiritists increasingly welcomed more skeptical observers into their circles.

Leymarie pursued his interest in developing a more scientific spiritism further. In 1878, he helped found the Société scientifique d'études psychologiques (SSEP) to promote the experimental study of spiritism. "For a long time [the phenomena were] presented as miraculous by superstition and, as such, dismissed by science. Now, we know that facts of this kind are perfectly natural, that they have their laws like all beings and all forces have their laws," he stated in the mission and objectives of the new society.[71] The aim of the SSEP was to provide spiritist phenomena with a scientific basis through the study and control of their production. It was not solely limited to an experimental approach and also promoted reflection on the moral consequences of a more rigorous spiritism, with its demonstration of a future life and communication between the living and the dead: "It is nothing less than the starting point of a new social order and the promise of a better humanity."[72] Initially, the work of the society was published in the *Revue spirite*, but, starting in 1882, the SSEP published its own bulletin for a short while. Until 1883, it offered weekly lectures on philosophical and psychological questions related to séances. Although short-lived, the SSEP does suggest a significant shift in the spiritist movement at the time. Spiritism had taken a turn toward physical manifestations. Spiritists now encouraged the presence of a scientific outlook alongside their doctrine. In 1888, this trend became more apparent in the *Revue spirite* itself, which changed its subtitle from the original "*Containing the story of material or intelligent manifestations of the spirits, apparitions, evocations, etc., as well as all the related news in spiritism. — The teaching of spirits on the things of the visible and the invisible world, of the sciences, morality, immortality of the soul, the nature of man and his future. — The story of spiritism in antiquity; its relation to magnetism and somnambulism; the explanation of legends and popular beliefs, of the mythology of every people, etc.*" to the more concise "*Journal of psychological study and experimental spiritualism. Bi-monthly magazine founded in 1858 by Allan Kardec.*"[73]

The first two decades after Kardec's death were also marked by a new interest in séances from outside the spiritist movement, particularly Great Britain, where spiritualism was never as unified a movement. From the beginning, British scientists had had a much greater access to séances, and, unlike the early French séances, which had been dominated by automatic writing, British séances tended to be filled with physical phenomena and thus more conducive to scientific observation and experiment. In 1869, the same year Kardec died, a report by the London Dialectical Society launched the field of psychical research. In the early 1870s, the chemist William Crookes experimented with the mediums Daniel Douglas Homes and Florence Cook. His observations of their séances became landmarks for the embryonic discipline and helped to sustain interest enough to create the Society for Psychical Research in 1882.

In France, repercussions of all of this novel work were clearly felt as a few more or less successful ventures began to surface. In 1874, the *Revue expérimentale de psychologie*, a journal dedicated to the development of a psychology of the soul, began a two-year run. It featured articles on sleep, somnambulism, hypnotism, and spiritualism and devoted a significant place to British psychical research, bringing it to the French public. A combination of reports and comments on British psychical research, and articles on somnambulism and hypnosis gave the journal a distinctively French flavor. In 1875, the publisher of the journal, the botanist Timothée Puel, wrote that after having given much space to British psychical research in its first year of publication, the *Revue expérimentale de psychologie* would now, it was hoped, be able to show that France was not lagging as far behind as it was believed when it came to such work. He also mentioned the recent creation of a French committee consisting of physicians dedicated to the scientific study of physical and psychological manifestations of the marvelous. The committee would reflect on means to centralize the efforts of isolated individuals or groups for a better understanding of these phenomena. The *Revue expérimentale de psychologie*, it was promised, would continue to bring news of the French committee and publish its proceedings as soon as its first meeting was held.[74] Unfortunately, this committee was not mentioned again, and shortly afterward, in 1876, the *Revue expérimentale de psychologie* ceased to appear.

In 1885, the Société de psychologie physiologique was created, with a clear interest in séances. Its stated agenda was the investigation of the subconscious mind. The list of its members was impressive and included Pierre Janet, Théodore Ribot, and Charles Richet, as well as the eminent Jean-Martin Charcot of the La Salpêtrière hospital as its president. The focus of research was never spiritist phenomena, but the topic was occasionally brought up in the pages of the journal, albeit always in physiological and pathological terms. Although the society did not survive Charcot's death in 1893, its existence was indicative of the increasing attention that spiritism was now attracting in some established scientific circles.[75]

If the 1870s and 1880s did not witness the successful establishment of a tradition of psychical research in France as they did in Britain, Kardec's death in 1869 did allow a more scientifically oriented spiritism to flourish and provided a limited space for new societies and journals to emerge. Among these were the Union spirite française and the *Revue scientifique et morale du spiritisme*, both created by the engineer Gabriel Delanne. Delanne had grown up with séances. Born in 1857, the year Kardec published his *Livre des esprits*, he was part of the new generation of spiritists. His parents, both salespersons, had been part of Kardec's group and friends of the leader. His mother had also been a medium. In 1882, Delanne founded the Union spirite française with his father. He later created the *Revue scientifique et morale du spiritisme*, which ran from 1896 to 1914 and again from 1921 to 1926. At the time, the journal presented the most scientifically minded version of the spiritist doctrine. Delanne's spiritism connected the spiritual, the philosophical, and the rational.[76] He believed that spiritism was a science that would bring proof of both the existence of the soul and immortality through a dialogue with the spirits.[77] He differed from Kardec in his wish to bring spiritism into the realm of science; and he differed from Flammarion in that he still affirmed the role of the spirits at séances.[78] For Delanne, the time had come to introduce the public to a new brand of spiritism, one that would not be speculative but based on facts. "The present generation is tired of metaphysical speculations," he wrote; "it refuses to believe what has not been absolutely proven."[79] Spiritism would be a science of both revelations and observations.[80] After all, spiritism and science were similar in that they both progressed gradually: science through observation and experiment, and spiritism through observation, experiments, and revelations. As such, the doctrine was not set in stone and would evolve along with our intellectual development.[81]

Like Kardec, Delanne believed that the key to obtaining definite proof of the afterlife lay in the concept of the *périsprit*. In the *Livre des esprits*, Kardec had explained that humans were composed of three principles: the material body, which would decompose at death, return to the earth, and complete the material cycle of life; the immortal soul imprisoned in the body for a lifetime; and the *périsprit*, which joined body and soul during life and escaped with the soul at death. Although the *périsprit* remained attached to the body during life, the bond between body and *périsprit* could be weakened. At times and for a short period, the *périsprit*, and, along with it the soul, could leave the body and travel to other locations.[82] It was their *périsprits* that allowed spirits to materialize themselves or diverse objects at séances. Delanne saw great potential in the notion of the *périsprit*, because, in its physical manifestations, it could be studied scientifically. It was at the heart of his quest for a proof of reincarnation. The spirits had revealed that the *périsprit* came from the universal fluid of our planet, a primitive matter from which all bodies came through successive transformations. This, Delanne wrote, could be studied and proven, all that was needed was to demonstrate that "matter can exist in different states, each one simpler than the previous," and "that the infinite variety of bodies can be expressed in this unique matter."[83]

For the spiritist, if these two propositions were shown to be true, the existence of a universal fluid would be proven. With this, he argued, spiritists would be one step closer to proving the existence of the *périsprit* and of reincarnation. Delanne hoped that notions of the *périsprit* would soon begin to surface in the biological sciences. After all, he wrote, it was a powerful explanatory tool that could account for both mental and physical heredity better than any other scientific theory: the *périsprit*'s capacity to organize matter explained physical heredity, while the existence of a reincarnated soul attached to the body by the *périsprit* accounted for a continuity from one incarnation to the next, thus explaining differences in intelligence and personality.[84]

A few others discussed the possibility of finding a proof of the afterlife in spiritism. Émile Boirac, rector of the Académie de Dijon, and thus in charge in matters of education for the region, for example, argued that séances opened the doors to the spirit world, which functioned according to laws discoverable by science.[85] Once determined, these laws would allow scientists to control the phenomena and reproduce them at will. Boirac pleaded with scientists to consider spiritism. Scientists, of course, should require experimental proof from a hypothesis, but they should never refuse to consider it, he wrote.[86] The convinced and enthusiast spiritist J. Malosse described spiritism as a science of the soul that would lead to the discovery of a unity in nature and a law of evolution connecting humans to God.[87] In a work recognized by the Académie des sciences, Léon Chevreuil argued that spiritism would lead to a synthesis of reason, certainty, and faith.[88] Although he did not consider himself to be a believer, Chevreuil thought that the existence of a soul that escaped the body after death to survive had been proven by science. Moreover, a number of facts, this time much more difficult to observe, showed that the dead could sometimes manifest themselves in the world of the living.[89] Other authors of the time also appealed to science to build a theory of the afterlife. The philosopher Pierre-Camille Revel, for example, proposed a theory of metempsychosis based on biological knowledge of the period. This posited that at death, a gas released from the body combined with the surrounding cosmic dust and organic matter and finally fixed itself and developed into a being of the same species. This being, Revel claimed, would have a new personality but still remain the same person.[90] Whatever their justification for it, most writers who advocated a scientific theory of the afterlife used the phenomena of séances to argue their case.

In 1912, Gustave Geley, a physician who would spend a lifetime researching mediumistic phenomena wrote a short text on reincarnation. Commenting on the moral, philosophical, and scientific implications of his beliefs, Geley drew an interesting picture of the doctrine. Like Delanne, Geley believed that including reincarnation in moral and scientific considerations would produce a more complete picture. Reincarnation, he wrote, is based in an immanent justice. All acts have their fatal repercussion and reaction in one or another existence. The doctrine thus takes away the need for divine judgment or supernatural sanctions. For Geley, reincarnation also explained some philosophical problems. Evil became a measure of the inferiority of worlds, not a reality in the absolute sense. The doctrine of reincarnation was consistent with the realization that there was nothing special about the Earth or humans. Geley also argued that a belief in reincarnation was in agreement with the scientific knowledge of his time in astronomy, natural history, geology, paleontology, anatomy, comparative physiology, and evolution. Reincarnation, according to Geley, even helped to refine notions of evolution, because "unknown factors in evolution" would be revealed by the study of the soul. It could bring a greater understanding of talent, genius, and the enormous differences between physical and psychical heredity, for example.[91] A belief in reincarnation was compatible with research on hypnotism, somnambulism, and mediumistic phenomena as well. As a hypothesis, reincarnation was in agreement with modern science and solved many philosophical and moral problems, but it had yet to be proven through conclusive research. Geley hoped that his own work on mediumistic phenomena would help to validate the doctrine.

∽∾

In the first few years of their existence in France, séances provoked fleeting discussions in every corner of society. For those with an interest in spirituality and the supernatural, they inspired a more sustained interest. After 1857, those following the spiritist doctrine came to believe that the moving and talking tables were manifestations of the spirits. They believed that the phenomena provided a direct contact with the world of the dead. For spiritists, séances offered a whole lot more than direct access to religious and moral knowledge. They provided a space in which tangible confirmation of the existence of an afterlife could be obtained, a site where the supernatural could be witnessed and, to a certain extent, controlled and reproduced. For believers, Kardec's work was thus invaluable. More than the simple establishment of a doctrinal tradition, it taught spiritists what counted as a phenomenon, what procedures and questions were valid, which facts could be accepted and which should be rejected. It taught them to be witnesses and experiment on a set of phenomena in their own societies and homes. But spiritism was just one way to produce, understand, and interpret the phenomena witnessed at séances. By the late 1880s, spiritists were beginning to encounter other groups who shared an interest in the manifestations of the spirits with them but did not agree with their explanations. In particular, occultists, members of an unorthodox group growing in importance in France as elsewhere, were now adapting and redefining the phenomena of the séance to fit their own interests and beliefs about the supernatural.

CHAPTER TWO

Occult Wisdoms, Astral Bodies, and Human Fluids

In September 1889, as many as forty thousand participants came to Paris for the first meeting of the Congrès spirite et spiritualiste international. Although united in their belief in the authenticity of mediumistic phenomena, those who partook in the proceedings adhered to no set explanation about them. At the congress, Allan Kardec's followers were confronted with a diverse array of theories and doctrines on the manifestations of séances. The world of French spiritism was widening to include new phenomena and competing discourses. The British and American spiritualists present at the meeting accepted the reality of spirit communications but did not necessarily adhere to Kardec's doctrine. As for the delegates who identified themselves as occultists, they diverged from French spiritists in more significant ways. Theirs was a different set of interests. They even refused to accept the spiritists' most fundamental tenet: the role of spirits at a séance. In their view, mediums and participants were causing the phenomena, not supernatural beings. For occultists, the importance of the manifestations of séances lay, not in the hope of proving the immortality of the soul, but in the support they could bring to ancient revelations and to uncovering long-forgotten human abilities. Spiritists were not necessarily opposed to such teachings. For most of the 1880s, the *Revue spirite* had been trading Kardec's automatic writings and Leymarie's physical productions for a more occultist orientation. Throughout the decade, the French occultist revival had been gaining strength. By 1889, the prominence occultists were given at the spiritist congress left no doubt as to the position they had attained in unorthodox scientific circles in France.

French occultism was never a unified movement. Unlike spiritism, which was centered on a single doctrine, it developed around a series of more or less success-ful leaders and journals. There were many schools of occultism, some associated with kabbalism, others with various hermetical traditions of the Renaissance, and yet others with various so-called lost traditions of the East. No matter what their particular brand of occultism, however, all the occultist schools shared in the same method of teaching and the same general objective. Occultist knowledge was always esoteric knowledge. It was to remain hidden from the general public and to be revealed only to a select few through a series of initiations and rituals. Nineteenth-century occultism was also understood to be much more than a cry for the return to ancient wisdom and forgotten ways of knowing. Its supporters hoped that it would bring about the dawn of a new science. What occultists meant by science, however, had little to do with the methods and questions pursued in universities, research establishments, and learned academies around the country. They used the word loosely, playing on the prestige and the authority of a field, while rejecting most of its content and approaches. They saw themselves as work-ing toward a new conception of science, one that would incorporate and indeed give prominence to ancient knowledge and esoteric research methods. Observa-tion and experimentation would still have their place in such a science, but they would be given a supporting role, subordinate to wisdom and revelations. Some occultists had a scientific background, mostly in medicine. They hoped to un-cover supposedly lost knowledge embodied in ancient alchemical, astrological, magical, and esoteric traditions and to reconcile those traditions with the modern corpus of the physical and physiological sciences. Even for them, however, the concept of an occult science remained fluid and malleable. Like others in occult-ist circles, they believed that authority lay first and foremost in the sacred revela-tions of the ancients and only later in observation and experimentation.

Although never a significant group in terms of the number of its adherents, occultism had a profound impact on the fine arts, music, and literature of the late nineteenth century. Its influence was evident in a number of artistic currents, most notably the decadent and aesthetic movements.[1] Occultists denied the im-portance of rationality, preferring instead to focus on sensations and instinct. They were fascinated by symbolism and saw hidden signs and secret meanings every-where. In their ways of understanding nature, they expressed their disapproval with the wider culture of their era. Those more scientifically oriented in the move-ment deplored the materialist attitude they felt was dominating the sciences at the time. They hoped that the rediscovery and reinterpretation of ancient wisdom alongside the inclusion of the phenomena of the séances would bring about a new era of scientific exploration, one that would be focused on the lost knowledge of the past and the impressive abilities of mediums and other mystical individuals of the present.

FRENCH OCCULTISM FROM LÉVI TO PAPUS

A renewed interest in all things occult developed in France in the mid nineteenth century, mostly fueled by the work of Alphonse-Louis Constant. Throughout the 1840s, Constant had been interested in Christian socialism and writing critical commentaries on biblical and other sacred texts. In 1853, shortly after his wife left him for another man, he experienced a spiritual crisis of his own and turned to the occult. He adopted a Hebraized version of his name and became Eliphas Lévi, writer of commentaries on the Tarot and magic. In an attempt to create a unified doctrine, Lévi fused different traditions of the occult into a new doctrine he labeled occultism. The term would win popularity in the 1870s. By the time he died in 1875, occultism had gained momentum both in France and abroad, not as a unified or coherent doctrine, but more as a guiding idea encompassing various teachings and traditions.[2]

This emergent interest in the occult occurred in the context of a growing fascination with the East. In France, the exoticism and aura of mystery associated with distant lands inspired many prominent men and women. Novels on the topic abounded. Explorers reported on their travels and encounters with strange and exotic populations. At the beginning of the twentieth century, for example, the explorer Alexandra David-Néel gained fame through her travel accounts and descriptions of Eastern mysteries. Living in and traveling through India, Japan, and later on China and the Soviet Union, David-Néel experienced and wrote about beautiful landscapes, strange customs, and deep spirituality. She also told tales of her conversion to and understandings of Buddhism, her meetings with the Dalai Lama, and even her visit to Lhasa, a city forbidden to foreign visitors at the time, all of which enticed readers back home.[3] Other authors focused on the magical and wondrous mysteries of faraway lands. Tales of fakirs—indigent mystics detached from materialistic values, living above earthly concerns—and their amazing abilities were popular at the time. They could reportedly climb up ropes hanging in midair and control vegetable growth. Some were said to have the ability to reduce all substances into undifferentiated atoms. Others were believed to be shape-shifters or to be capable of either self-levitation or levitation of objects. Still others could cure through their vital fluid, command other beings, or foresee future events, or events taking place in other astral realms.[4]

Occultists fueled this interest in the mysteries of the East. Some, like Paul

Gibier, believed that ancient teachings had been preserved by an elite priesthood and still lay hidden in underground sanctuaries protected from floods and other disasters at high altitudes inside the mountains of Asia.[5] Others tried to organize teachings around ancient Eastern wisdom. In 1875, the same year Lévi died, a new occultist movement building on this enthusiasm for things Eastern was created in New York when Helena Petrovna Blavatsky founded the Theosophical Society with the help of the lawyer and journalist Henry Steel Olcott. Theosophy consisted of a set of mystical teachings inspired by esoteric traditions of the East. In its aims, the Theosophical Society demarcated itself from other occultist groups by its discussion of universal fraternity and its focus on the study of a set of Aryan and Eastern teachings. Not concealing a certain derogatory attitude toward the contemporary populations of the East, theosophists acclaimed the glorious character of ancient Eastern civilizations, but believed that the sacred wisdom of those times had been distorted and lost through years of decline. Thus, a straightforward study of contemporary Eastern beliefs was not sufficient and would only lead to popular and superstitious versions of what had once been an extraordinary and rich knowledge. To uncover the wisdom of the past, theosophists felt they needed to reconstitute and extrapolate from current traditions and beliefs.

Within a few years, the theosophical movement began to consolidate itself. In 1877, Blavatsky's book *Isis Unveiled* provided a clearer doctrine for the movement. In 1879, the headquarters of the society moved from New York to India. There, Olcott and Blavatsky acquired a house in Adyar, near Madras, and established the society where it remains to this day.[6] Blavatsky left India for London in 1887, but Olcott stayed on to recruit theosophists, teaching them a mixture of Buddhism, theosophy, and other traditions. As for Blavatsky, she continued to promote the society's interests and the theosophist doctrine through lectures, journals, and books. *The Secret Doctrine*, her most popular work, appeared in 1888. Meant as a synthesis of theosophy, it was hailed by her supporters as a triumphant effort to uncover and expose an elemental tradition. For critics, however, it was yet another confused attempt at combining different ancient doctrines and traditions, which it assembled into a disjointed whole.[7]

Theosophists hoped to attain truth and enlightenment on questions of life and death. They were not experimentalists. Theosophy was a doctrine that would provide an answer to their quest: "[Experiment] does not intervene to orient and direct research. It intervenes to justify, if it has to, the positions adopted, in immemorial times, with regards to the secret doctrine," one of them wrote.[8] As such, theosophists were never as practically inclined as other occult groups would be. Like all occultists, however, they adhered to the revelations found in ancient texts, both existing and lost. Their literature was filled with references to the lost world of Atlantis and the great knowledge possessed by its elite, which had been transmitted to the ancients of India and Tibet before its catastrophic destruction. Centuries later, the sacred doctrines now lay hidden and buried under layers upon layers of Western Judeo-Christian religious traditions and beliefs. For theosophists, one of their vital tasks was thus to uncover the sacred knowledge by deciphering its remaining relics.[9]

Theosophy was never as popular in France as in the Anglo-Saxon world, but it introduced many to occultist traditions and functioned as a catalyst to the larger movement. The first attempt to establish a French theosophical tradition was made in the mid 1880s. At the time, Lady Caithness (Maria de Mariategui), a friend of Blavatsky's, conducted a fashionable salon frequented by many occultist sympathizers. In 1884, she created the Société théosophique d'orient et d'occident in Paris, and in 1886, she launched L'aurore, the first journal of theosophy in France, which published reports on English theosophical research, adding its own Roman Catholic flavor to the mix.[10] Lady Caithness's journal lasted for only a few issues, but theosophists continued to try to establish their doctrine in France, albeit without much success. By 1887, L'isis, a French branch of the Theosophical Society, was created. Its journal, Le lotus, which featured mostly translations of English works, appeared for only two years, after which it was replaced by the Revue théosophique, which ran for one year as the official organ of the Theosophical Society in France. "What we have hope to reveal to the reader with the creation of this journal," wrote its editor, the comtesse d'Ahémar, "is a science that is as old as the world and still new to our Western world, as rich as it is when it comes to science."[11] The secrets of an ancient science, those that remained buried in Tibet and India, were waiting to be discovered by a French theosophical movement. The Revue théosophique would provide a comprehensive presentation of this developing science for the French public. Here again, the journal showcased translations of Blavatsky's and other writings in English on theosophy. In 1891, the comtesse d'Adhémar abandoned the journal for personal reasons, and it was quickly replaced by Le lotus bleu, edited by Jean Matthéus, which survived until 1986.

For most of the 1880s, the occultist revival taking a hold of France centered on the theosophical doctrine. By the end of the decade, however, amid quarrels and disagreements among editors and contributors, the movement began to organize itself around a new set of journals, societies, and charismatic and colorful leaders. The poet Stanislas de Guaïta, for example, renowned as much for his work on the sciences maudites as for his excessive consumption of morphine and opium, left behind voluminous writings on mystical and Christian occultism. In 1889, Guaïta created the Ordre kabbalistique de la Rose-Croix, hoping to bring about a French renaissance of Rosicrucianism, a German mystical movement claiming roots as far back as the Middle Ages. In 1896, together with François Jollivet-Castelot, Guaïta founded the Société alchimique de France and reorganized L'hyperchimie, a journal that had existed since 1875, to promote the revival of the art of alchemy in France. Guaïta was surrounded by controversy and quarrels, including a "war of the magicians." In 1893, the writer Joris-Karl Huysmans, an important literary figure, accused Guaïta and his order of having provoked the sudden death of a former priest turned satanist, Joseph-Antoine Boullan, through magical means. Boullan had struck up a friendship with Huysmans while the latter was working on La-bas, a novel depicting the occult underworld of the capital, with its satanic masses and black magic. The writer and occultist Jules Blois, another of Boullan's friends, challenged Guaïta to a duel, which never took place.[12]

Guaïta and Huysmans did much to associate occultism with satanism in the public's eye, and stories of satanic ceremonies and black magic began to captivate readers of the popular press. "Dr Bataille," the anonymous author of a weekly feuilleton titled Le diable au XIXème siècle, for example, told of satanic rituals he had supposedly witnessed in his travels around the world, notably of dealings with Satan at a certain Masonic lodge in Charleston, South Carolina. For two years, the French public read about these American devil worshippers with interest. They learned of Sophia Walder and Diana Vaughan, who were supposedly both members of the lodge and lovers of Satan. A prediction was made that Sophia would soon become pregnant by a demon named Bitru, that the child born of this union would see the light of day in Jerusalem, and would one day beget the Antichrist. In 1895, Diana, now converted to Catholicism, allegedly wrote her memoirs, recounting her past experiences back in Charleston with Satan and a demon named Asmodi. The memoirs seemed to confirm the popular French feuilleton's chronicle and were taken as a confirmation of the story. The tale continued to unfold in incredible ways and in the public eye for a few more months but was later revealed to have been a complete hoax produced by the anti-Catholic polemicist Léo Taxil.[13]

Some occultists played upon the fears instilled in the public about their practices. Joséphin Péladan, another colorful literary figure of the movement, often dressed as a priest of the dark forces and enjoyed playing on ambiguities of good and evil. His work focused on the esoteric traditions he believed to be buried within Christian teachings. Initially a member of Guaïta's order, Péladan left the group in 1890 to form his own dissident order, the Rose-Croix catholique, and

pursue his search for traces of an elusive lost science within the Rosicrucian traditions.[14] Not all occultists enjoyed the sensationalism and controversies associated
with their movement, however. Some worked hard to separate themselves from
this more prurient association with satanic rituals and called for greater attention
to be paid to a nobler goal of the movement, the development of an occultist
science. Fearing that it would deflect the attention away from the scientific part
of the movement, the occultist Ernest Bosc even asked the Catholic clergy to re-
frain from making accusations and from promoting a vision of occultism that was
strongly associated with satanism.[15]

Bosc was not alone in his focus on an occultist science. In fact, one of the leading figures of the movement in France was predominantly interested in the kinds
of contributions that occultism could make to whole of the sciences. Like many
others, Papus, a physician by training, born Gérard Encausse, had encountered
the movement through the French Theosophical Society. In 1888, he left the
society to create his own journal, L'initiation, and, a year later, founded his own
society, the Ordre martiniste. His decision to leave the theosophical movement was
associated with his belief that occultism should be based on the Judeo-Christian
tradition and the teachings of ancient Egypt, rather than on the lost traditions of
the Tibet. Papus had discovered occultism through the work of Eliphas Lévi and
adopted his occult name from one of his predecessor's books. His particular brand
of occultism came not only from Lévi but also from Fabre d'Olivet, an eighteenth-
century poet and biblical commentator with an interest in Pythagoras, as well as
from Maître Philippe, a spiritual healer who was Papus's teacher. Papus claimed to
have learned the symbolism attached to occult knowledge from Lévi and d'Olivet.
He had discovered Jewish mysticism and the esoteric traditions buried within
Christianity from their work. From Maître Philippe, he had adopted the notion
of progress through multiple existences and the search for peace as a sign of that
progress.

Papus had an ambitious program. He hoped to bring about the spiritual renaissance of his times and unify all occultists under a single institution. He would
uncover the basic esoteric tradition common to all religions. By blending ancient
wisdom with the observational and experimental knowledge found in contemporary science, he would provide a synthesis of all knowledge. As such, he would
successfully unify the occult with the visible and the metaphysical with the physical. With his new philosophy based on ancient teachings, he would rid the world
of scientism, supply those who needed it with a weapon against materialism, and
stand against militarism and misery. To this end, Papus created the Ordre martiniste, "a school of moral chivalry" with the aim of developing spirituality in its

members through devotional exercises, the study of the unknown, and the creation of a scientific faith based on observation.[16]

The society functioned as an initiation site. Not unlike scientific education, occultism required intellectual work from its adherents in order to master its concepts, but, as with all esoteric teachings, occultist learning was based on initiation. Members would be secretly and slowly introduced to the sacred wisdom.[17] A tradition said to be rooted in ancient times, initiation consisted in the gradual education of a scholar from an initial dogma to the development of a set of transcendental faculties and the final "perception of Principles of the Infinite and Absolute, otherwise inaccessible" to him.[18] The learner began by receiving the "dogmatic and succinct teachings of a great synthesis" and later worked with his masters to acquire the hidden knowledge of the occult.[19] Occultist learning was metaphysical in nature. The knowledge of the initiate was not in competition with contemporary scientific discoveries. It was understood to exist on the margins of science, ready to provide a new framework of understanding that would combine observation, experimentation, and initiation.[20]

The Ordre martiniste was grounded in a tradition of Freemasonry and secret initiations. It took its name from the eighteenth-century mystic Louis-Claude de Saint-Martin, who was said to have been inspired himself by the work of Martinez de Pasqually, a mystical Freemason of the first half of the eighteenth century. Saint-Martin had preached that humans needed to reclaim their glorious past as androgynous creatures capable of commanding the spirits, a position they had lost with the original sin. His teachings had supposedly been secretly transmitted from generation to generation. The origins of the Ordre martiniste were said to rest in the unification of the knowledge of Saint-Martin possessed by Papus and by his colleague and fellow occultist Pierre-Augustin Chaboseau, each man having been initiated to the teachings of Saint-Martin separately.

Beyond its teachings, the order's organization, with its supreme council of twelve members and its different steps and degrees of enlightenment, was also inspired by Freemasonry.[21] At meetings, members were said to gather in a spirit of altruism opposed to the materialism and egoism that plagued contemporary French society. Groups were established to study what was described as marginal and rejected science, a mysterious and ancient knowledge complemented by contemporary discoveries.[22] It was hoped that the Ordre martiniste would bring enlightenment to those who had lost faith in the modern world. It was a success. Throughout the 1890s, the Order established itself not just in France but across Europe, reaching Russia and the court of the Czar Nicholas II by the end of the decade. At the Russian court, the order's popularity grew. By 1905, faced with significant social and political tensions, the czar even summoned Papus to St. Petersburg. Evoking the spirit of the previous czar, Alexander III, Papus is said to have recommended repression as a necessary measure and announced the coming of a great revolution within the empire. Although Papus's influence at the Russian court is said to have been significant for a while, it was rapidly eclipsed by the arrival on the scene of another mystic, Grigori Rasputin. As for the Ordre martiniste, it flourished until the death of its founder in 1916, continuing to recruit members from all over Europe.[23]

Often criticized by fellow occultists for his more lenient attitude toward initiation and secrecy, Papus tried to reach the larger public through a series of books and his journal, L'initiation, which ran from 1888 to 1912. Believing that everyone should be instructed in occultist teachings, Papus organized L'initiation didactically in three sections: the first introduced occultism and provided an initial guide to those interested in understanding the esoteric revelations buried under complicated and meaningful religious and spiritual symbols; the second was devoted to the philosophies and knowledge associated with occultism; and, finally, the third was dedicated to literature and poetry, presenting occultism in a lighter manner, for what Papus imagined would be a feminine audience.[24]

In his mission to educate, Papus also created the École supérieure libre des sciences hermétiques de Paris, which taught its attendees "academic" occultism. The school was designed to accept around thirty students each year. The venture seems to have had a certain success. By 1901, courses were moved from their initial small room to a more spacious setting with two classrooms. The instruction and curriculum was divided into three stages. First, students were introduced to an occult synthesis of ancient revelations and their adaptation to nature and contemporary science. After three terms, more specific occultist theories were presented to a group of carefully selected students who had gone through a screening process based on their invisible auras. At this stage, Papus was hoping to have eliminated all those who had only been interested in gaining impressive power, and to be able to concentrate on the initiation of the truly devoted students, whose interest in occultism was selfless and noble. In the final stage, courses dealing with practical matters and the realization of the doctrine were taught orally, because the esoteric character of higher occultist learning had to be respected.[25]

At the center of Papus's teachings was a belief in the ancient traditions of Egypt, with reincarnation as the central tenet.[26] Like Kardec and the spiritists, French occultists believed in reincarnation and saw it as the return of the spiritual being to a new physical envelope. This return could take place anywhere in the cosmos. As such, it unified all things in the universe and all planes of life.[27] In death, each principle that constituted the human body returned to what was believed in these teachings to be its own plane: bones and minerals returned to the mineral realm; muscles returned to the plant realm; and animal cells returned to the animal realm. While the physical body returned to the physical plane, the astral body and the spirits traveled to the spiritual or divine place until the time came for a new physical existence.[28] This new human life would be "the mathematical result of former existences," as illustrated by the Pythagorean Theorem, in which the two legs of a right-angled triangle were held to represent will and divine providence, while the hypotenuse represented destiny. Thus, for each being, the square of the will added to the square of divine providence equaled the square of destiny.[29]

Papus claimed that traces of these lost traditions could be found in Christianity. In his writings, Jesus and Moses were both associated in their messages and approaches with the ancient tradition of initiation. Moreover, Papus argued that references to the doctrine of reincarnation were present in the Bible in the revelations of both Eli and John the Baptist. Parables such as that of a man born blind being punished for his past sins could be found in many passages. Papus took these as proof that Christianity had been based on older Egyptian teachings and that the doctrine of reincarnation was as much a part of the secret teachings of the Church as the initiatory knowledge had been a part of the sacred knowledge of an ancient Egyptian elite: "The biblical teachings on reincarnation are made up, as always in sacred matters, of a part officially communicated to the population and of a complement secretly communicated to the masters."[30] Reincarnation could be easily reconciled even with the official teachings of Christianity if the lapse of time between judgment after death and the last judgment was interpreted as a moment during which Heaven, Hell, or Purgatory could be experienced in a materialized form.[31]

Papus summarized esoteric knowledge in three points:

1. The existence *of the Tri-Unity* as a fundamental law of action in all planes of the universe.
2. The existence of *Correspondences* intimately uniting all the visible and invisible portions of the universe.
3. The existence of an *invisible world*, exact double and perpetual element of the visible world.[32]

The notion of a unity of all things in the universe and a correspondence between man and nature were important aspects of Papus's teachings. He saw a correspondence between humans and the Earth, with the atmosphere as a breathing system. Humans were united to all other creatures and each movement in the universe by their unconscious.[33] The human body was constructed like the universe, with globules as suns around which small corpuscles circulated as planets.[34]

For the occultist and Russian statesman Maximilien de Meck, occultism held the promise of completing contemporary science and providing it with a renewed sense of spirituality. Together, the two could become a science of the invisible and the visible, a synthesis of all knowledge.[35] Papus himself divided occultism into two parts. One was tradition; it was immutable and found in the hermeneutic writings of all times and places. The other consisted in the personal research of an author through commentaries and applications.[36] It was in this latter part that he felt occultism and science would unite to form a total science, the union of a thesis (physics), an antithesis (metaphysics), and a synthesis (mathematics). In this way, the two halves that had been separated at some point in the distant past would be joined once again. The official knowledge and the secret knowledge, kept from generation to generation by fraternities or assemblies of the initiated, would complete each other and lead humanity in its search for Truth.[37] Papus was often criticized by fellow occultists for developing a less spiritual occultism by introducing modern concepts of experiment and observation into the movement.

Inspired by the work of Paracelsus and other Renaissance scholars, nineteenth-century occultists turned to magic, divination, astrology, and alchemy. They described a world in which all things were unified; and this unification in nature would become visible through the study of magic. Whereas Lévi had taken a more mystical approach to the art of magic, Papus gave it an experimental flavor. He defined magic as the application of a human will to the forces of nature. As such, magic held the crown among all the biological sciences.[38] He saw it as central to his endeavor to unite ancient revelations and contemporary scientific observations and experiments. Magic would work as a connection between the occultist tradition and contemporary science, particularly physiology and psychology. It would bring about a total science of nature, the body, and the mind.[39]

In emphasizing the importance of magic and in presenting it as the basic development of occultism in the modern world, Papus was not alone. Joseph Maxwell, a lawyer in Bordeaux, saw magic as the active form of the religious sentiment. That magic was possible was proven by the fact that it had persisted over centuries as a useful way to understand and control nature. Maxwell did not, however, think the ancient esoteric teachings on magic revealed sacred truths. He argued that magic was a relative phenomenon, that it was a temporary means to explain the world, complementing religion.[40] As science had progressed, magical facts had survived and continued to evolve, but the classification and the explanations given to different phenomena had changed and would continue to do so. With time, the older science of natural magic would be replaced by the new and emerging science of psychical research.[41] The seat of magical forces lay in the subconscious; they were the forces responsible for mediumistic phenomena. With the development of psychical research, Maxwell wrote, we would come to realize that magical abilities and mediumship were both made possible by forces outside our personalities and constituting our true individuality.[42] Similarly, the future of a science of divination lay in psychical research, because the ability to foresee certain events also emanated from a subconscious sensibility.[43] Occultists had to acknowledge that magical facts, like mediumistic phenomena and divination, had been observed for centuries, but they needed to reject ancient explanations of such phenomena and explore the new theoretical possibilities offered by psychical research. Such work, Maxwell believed, would lead to new theories of personality and a new occultism devoid of mystical qualities, which could then be incorporated into current science.

The belief that magic would eventually be incorporated into psychical research, and from there into science, was the typical argument interested psychical researchers used to counter occultist explanations. Fabius de Champville, president of the Société magnétique de France, asserted: "Occultism is neither a science nor a religion; it is the set of sciences still badly known or insufficiently defined."[44] Joseph Grasset, a physician who developed a theory of personality based on his observation of mediums, understood magic and occultism in a similar way. He distinguished between what he called the occultism of yesterday and the occultism of today. The occultism of yesterday focused on phenomena supposedly explained by the science of the day, such as animal magnetism and turning tables. The occultism of today involved the study of phenomena still unexplainable by science: spiritist phenomena, psychical radiation of the astral body, and telepathy. For Grasset, such phenomena would eventually move from the realm of occultism into science: "Science, which is never complete, everyday invades the domain of occultism of which the borders always move back and is thus like the *promised land* of science."[45] As science advanced, the frontiers of the mysterious and the supernatural would recede.

French occultism never became more than a marginal movement, but by the early twentieth century, its rival branches and conflicting theories had become part of the cultural landscape. Although its most prominent journal was probably *L'initiation*, a few other short-lived publications with similar aims were founded during this period. *La curiosité: Journal de l'occultisme scientifique*, for example, an attempt to launch occultism in the south of France, ran from 1889 to 1898 as an occultist journal, and later focused on psychical research. The *Écho du monde occulte* was published in 1905–6. *La haute-science* existed for a brief period in 1911, and ran articles on magic, astrology, alchemy, and mysticism. *L'écho du merveilleux* survived longest, running from 1897 to the Great War. Gaston Méry, its editor, declared it open to all philosophies but preferred an occultism that had a Roman Catholic flavor. *L'écho du merveilleux* concentrated on simple, direct reporting, bringing news of the events taking place in and around occultist groups. In 1908, a new society, the Centre ésotérique oriental de Paris, was founded, with *L'étoile d'Orient* as its journal.[46] If none of these ventures had the impact that *L'initiation* and its founder did, they nonetheless point to the persistent presence of a scientific outlook within the occultist groups of the period.

OCCULTISM AT SÉANCES

Until his death in 1869, Kardec had distanced his doctrine from esotericism. Although he had mentioned the ancient roots of spiritism, he had not elaborated upon this aspect of his teachings. Under his leadership, spiritism remained focused on present-day séances. By the 1880s, however, Kardec had been dead for more than a decade and a great deal had changed within the spiritist society. Among other things, Léon Denis, a spiritist with an interest in occultism, had come to the forefront of the movement. At the age of nineteen, Denis had discovered spiritism through Kardec's *Livre des esprits.* Living in the Loire Valley, he was only able to meet the leader of spiritism in 1867 during one of Kardec's promotional trips across the country. Throughout the 1870s, Denis had continued his spiritist activities and, by the 1880s, he had become a significant contributor to the *Revue spirite* and a central figure in the movement. In 1885, he was elected president of the Union spirite française. In 1889, he played an important role at the Congrès spirite et spiritualiste international.[47] That same year, he began a tour of France and Belgium promoting his spiritist beliefs. From the 1880s on, Denis's brand of spiritism came to be very influential within the movement, promoted by his contributions to the *Revue spirite* as well as his numerous books and lectures.

Through his work, Denis brought spiritism closer to occultism, ancient revelations, and esotericism. At times, his teachings echoed the work of Papus and other occultists more than that of Kardec. For Denis, the ancients had possessed great wisdom on matters of life and death. In the lost worlds of India, Egypt, and Gaul, sacred knowledge treasured and controlled by an elite priesthood had been divulged to the population and presented as magical religions. The scientific and philosophical character of such knowledge had been lost by those civilizations; only its religious side remained. Behind the symbolism and ceremonies used to capture the imagination of an impressionable populace, however, the secrets of an ancient elite remained buried. All mystical traditions, all religions, shared this hidden doctrine at their base.[48] Now, ready to be rediscovered once again, these ancient revelations would revitalize the modern world.

By integrating this lost knowledge, current science would widen its scope. Scientists might have gained insight into the world through the accumulation of facts and the deduction of laws, but they had so far kept their distance from superior principles and limited their search to causes and effects. In contrast, the elites of antiquity had been more concerned with the study of eternal principles. Denis believed that, through research on the marvelous, contemporary science had now arrived at the gates of this ancient knowledge, and, as scientists continued to explore these phenomena, a new and a more complete science would emerge.[49] An exploration of the supernatural would lead to the fusion of science and religion and the rise of a new doctrine, idealist in its tendencies and positivist and experimental in its method. From this union would come the resurrection of a secret doctrine, mother to all religions and philosophies.[50] Ambitiously, Denis hoped that knowledge of this secret doctrine, the lost wisdom of the ancients, would be incorporated into spiritism and give rise to a new society and an official state religion for the Third Republic. The age of religions as they had been conceived of until now was over, he claimed optimistically. The time had come for a new social and philosophical order.[51]

Denis emphasized that through occultism, spiritism would enable France to reconnect to its glorious past. Occultism would bring spiritism to new nationalist heights. Kardec had discussed the origins of spiritism in the religion of Gaul. For Denis, the celebrated past of a spiritist Gaul was further emphasized. The secret doctrine, he explained, had migrated from India and Egypt to France, where the druids had been introduced to it and had been able to use it to inspire courage in their warriors.[52] Denis believed that this brand of nationalism based on a common spiritual past and rationality would appeal to the French populace, and he tried to push spiritism in that direction. In the 1880s, under his influence, the Revue spirite adopted a more mystical flavor, allotting less space to automatic writing and the moral aspect of spiritism. The journal was now filled with references to witchcraft and occultism, accounts of goings-on in haunted houses, and articles on occult forces. Through the Revue spirite, spiritists were opening their world up to occultism.

Very quickly, occultists showed an interest in séances. If they did not follow Kardec and the spiritist doctrine, occultists did hope to find some proof of the veracity of their claims in mediumistic phenomena. For them, séances occupied a strategic position between the spiritual and the scientific, from which the fusion of occultism and science might emerge. As such, occultists were fated to interact with spiritists. At the 1889 congress in Paris, the two groups confronted each other. Spiritists believed in spirit intervention, while occultists did not appeal to the dead to explain séances. Nonetheless, they both agreed on the reality of the phenomena and felt united in a fight against those who doubted them.[53] The rhetoric of war against a common enemy was strong at the congress. In his opening comments, Jules Lermina, president of the occult section, emphasized the role of occultism in such a war: "This Congress is the battlefield of courageous intellects, defying the intolerance of those who seek to impose limits on the rights of analysis and investigation."[54] This was not a new idea for occultists, who often described their work as a struggle for Truth. Their battle cry appeared everywhere. L'initiation called for a unified front against materialism, the plague of the modern world, while L'étoile d'Orient gave hope to its readers by promising victory through perseverance. If not yet respected by scientists in their work, it promised that occultists would be hailed as the architects of a new and more complete science by future generations.[55]

For Papus, occultism did not contradict spiritism. It incorporated it into a more complete doctrine, one that called for a serious commitment to research and required some general understanding of modern physics, physiology, and psychology.[56] Whereas spiritists believed that all mediumistic communications or spirit manifestations came from the other realm, and distinguished phenomena according to different spirits, occultists were wary of such generalizations. They did not reject the idea that the dead could communicate with the living, but felt that a better explanation of séance phenomena than the spiritist one was needed. Various explanations were proposed. Trickery, human fluids, astral bodies, psychical forces, and spirits were each considered. Was the phenomenon caused by deception, by known physical forces, by some physiological force emanating from the medium, or by some psychical force? For occultists, only after all of these possible causes had been eliminated could the intervention of beings from another realm be considered.[57]

At the 1889 congress, Papus compared some typical explanations for various mediumistic phenomena given by spiritists and occultists respectively. For the occultist account, he used the concept of the astral body. Whereas Kardec and the spiritists believed mediums to be intermediaries between the living and the dead, Papus described them as subjects possessing the ability to leave their physical envelopes and travel with their astral bodies. Mediums might believe they were in contact with spirits, but, through their astral bodies, they were unconsciously

TABLE 1
Spiritist and Occultist Explanations of Séances According to Papus
at the 1889 Congrès spirite et spiritualiste international in Paris

	Spiritism	Occultism
The medium	Intermediary between the living and the spirits. Tool of the spirits in their diverse manifestations.	Being with a nervous system so constituted as to allow the astral body to get out very easily, acting unconsciously under the influence of the séance participants or the environment (physical or astral).
The table rappings	A spirit united with the medium's fluid acts upon the table.	The medium's astral body gets out unconsciously and lifts up the table, either alone or with the help of the astral bodies of the participants or of an elemental [spirit].
The table's answers	The conjured-up spirit is there and manifesting itself.	The unconscious (astral body) of a medium reads directly into the unconscious of a participant. Everything takes place entirely without the medium or the participants consciously knowing it.
Levitation of the table	Spirits lift the table.	The astral body of the medium united with the astral bodies of the participants produces this phenomenon. Elementals can participate in this.
Why darkness?	Spirits manifest themselves better and have more strength in the dark; hazy and diffuse light suits spiritist phenomena better.	Yellow light dissolves in the denser astral aggregations. Vital light that is invisible in light becomes visible in darkness.
The medium falls asleep	The spirits use the medium's fluid to produce the phenomena.	At the cataleptic stage, the emergence of the astral body is a lot more complete.
Little lights appear around the medium	It is the spirits that become visible through these phosphorescent lights.	The medium's life comes out through the spleen or the sympathetic plexus and becomes visible.

(continued)

TABLE 1
Continued

	Spiritism	Occultism
Objects are removed in the room	One of many material-ized but invisible spirits produces these phenomena.	The hands of the medium's astral body produce these phenomena from afar.
Fresh flowers fall on the participants	Spirits dematerialize the matter of objects and rematerialize it afterward.	The medium's astral body dematerializes the flowers, transforming them from a solid to a radiant state. With elementals and fluidic currents, the medium's astral body carries the flowers and suddenly re-materializes them in view of the séance participants.
Drawings and words appear suddenly	Spirits write or draw using the fluid they have at their disposal. The medium is the agent by which they manifest themselves.	The medium's unconscious writes or draws the images that float out of her/his mind or out of [the mind of] the participants. This action is caused by the medium's blood itself, which is materialized in black on the paper.
A material being appears. It talks, can be touched and photographed.	A spirit materializes itself by using every-thing that constitutes life, in the medium first, then in the participants and the surroundings.	1. The medium's astral body is united to an elemental and to the astral bodies of the participants. 2. This cluster takes the shape of the idea that dominates the medium or one of the participants. Mental sug-gestion determines the shape of the apparition. 3. This cluster has all the properties of material bodies.
The medium is awake and causes the materialized apparition	The spirit manifests itself to the medium. In this action, there are two distinct, con-crete individualities.	(We do not know the expla-nation of this phenomena according to occultism)
Molding and paraffin of the apparition	The spirit voluntarily produces this phe-nomenon in a way that must be very simple, but we do not know it.	Simple action of the medium's astral hand, which is dissolved after having been materialized.

TABLE 1
Continued

	Spiritism	Occultism
One of the participants, having broken the chain to grab the apparition, is hurt by the fall of an object.	The medium no longer has enough fluid to give the spirits, or the participants are no longer in mutual mental communion, the special state that is essential for the flawless execution of the phenomenon.	The magical chain, uniting all the astral bodies, created some sort of fluidic bed, on which objects were floating. The breaking of this chain brings the sudden fall of these objects.

Source: Papus (Gérard Encausse), "Les phénomènes magiques," *L'initiation*, April 1890: 17–22.

under the influence of their surroundings and the participants. This escape from their bodies explained why mediums seemed to know so much about their observers. The information was not obtained by a spirit, but through a connection between the medium's astral body and the spectators.

Mediumistic phenomena were of particular interest to occultists who, like Papus with the astral body, attempted their own set of explanations. The engineer and occultist Donald MacNab believed that spiritists focused far too much attention on spirit communication and thus encouraged mediums to describe their experiences in terms of another realm. He formulated an explanation of mediumistic abilities in terms of a yet unknown faculty of our consciousness. Mediums were psychic mirrors, showing the externalization of thoughts and the materialization of mental images in the participants. Mediumship revealed hidden parts of the unconscious. If mediums talked of spirits and the dead, it was only because they had been taught to interpret their experiences in such a way.[58] Like him, Blavatsky saw mediumistic abilities as significant. In her work, they became a sign of future human progress. One had to remain cautious, however. Impressive as they were, such powers could become dangerous when used without any theosophical understanding, she warned.[59] This belief that mediums and spiritists were manipulating dangerous forces and lacking the knowledge to do so was a common one among occultists at the time, especially theosophists.

As for spiritists, even they sometimes appealed to more occult concepts to explain the phenomena they witnessed. For example, Daniel Metzger, a spiritist whose lectures were announced and often discussed in the *Revue spirite* in the 1880s, believed that mediumistic abilities were explained by the particular molecular state of gifted subjects. He suggested that mediums were individuals in which the space between molecules was larger than in other individuals. The porosity that resulted rendered them more sensitive and better conductors of fluids. At séances, mediums would produce certain physical phenomena because of their capacity to use fluids.[60] The intellectual phenomena could also be explained in physiological terms as the product of the medium's ability to penetrate the "fluidic atmospheres" of others: "Every one of us is surrounded by a particular fluidic atmosphere. The medium, because of his great sensitivity, can easily penetrate the sphere of influence of those with whom he is put in contact," Metzger claimed.[61] Like Denis, Metzger reveals the importance that occultist thought had acquired even in the spiritist circles.

MEASURING FLUIDIC VIBRATIONS

Many occultists proposed the existence of a human fluid, a vague theoretical entity that could not only provide an explanation for mediumistic phenomena but account for other supernatural phenomena as well. Discussions of human fluids of all kinds were not new. A century before the advent of spiritism, the Austrian physician Franz Anton Mesmer had postulated the existence of a universal fluid uniting the whole of nature. Mesmer had believed that sickness arose from an imbalance of this fluid in the body. Animal magnetism was the method he developed for reestablishing balance in his patients through the touch of his hands. In 1778, after his practices were condemned in Vienna, Mesmer escaped to Paris, where he developed his theory further to include group therapeutics. There, he claimed that the universal fluid could become much stronger in a group. His new healing technique was a great favorite of the Parisian salons, where it resulted in strange but entertaining behavior by his patients, usually women, who convulsed and moaned in tanks of water, watched by audiences. Not surprisingly, such performances made physicians and scientists uneasy. In 1784, in an attempt to discredit Mesmer's practices, commissions at both the Académie des sciences and the Société royale de médecine were created to study the phenomena. It was imagination, and not a universal fluid, they concluded, that was the source of Mesmer's cures. Attacked by scientists, ridiculed in the press, and opposed by some of his disciples, Mesmer fled Paris and abandoned his research.[62]

The theory of animal magnetism did not die down with the departure of its creator, however. Throughout the nineteenth century, it continued to attract some interest in France, and a few journals remained dedicated to its practices.[63] Some, like J.-P.-F. Deleuze, the abbé Faria, and A.-M.-J. de Puységur, worked with and elaborated on Mesmer's notion of a magnetic fluid. Animal magnetizers thought that it was only a matter of time before the scientific community accepted the existence of this elusive fluid responsible for their phenomena. The notion of a universal fluid was compatible with magnetic and electric theories in the physical sciences, such enthusiasts were always quick to argue. Ether, the supposed gas or fluid in which phenomena occurred, was also frequently referred to in support of the existence of this universal fluid. Magnetizers were often convinced that understanding of fluids would soon enter the life sciences as well.[64] After all, vitalism, a set of theories stating that the difference between living and non-living matter resided in organization, special vital forces, and a chemical combination of matter, had become popular in the life sciences during the first half of the nineteenth century. Animal magnetizers often promoted the concept of the "Odic force," or Od, a supposed "life fluid" or radiation emanating from the sun, which was said to be perceivable by impressionable individuals. This notion derived from the theories of the noted German scientist Baron Karl von Reichenbach (1788–1869), a member of the Prussian Academy of Sciences, who had presented the Od as a key biological concept.[65] For magnetizers, it was a powerful explanatory concept. With vitalism and the Od in the life sciences and ether and the electric and magnetic fluids in the physical sciences, they were confident that their branch of learning would eventually take its place among the sciences as part of a new and comprehensive fluidic theory.

Thus, when séances began to be held in France, animal magnetizers had an explanation for what transpired at them: moving tables were a natural phenomenon caused by the same fluid that was responsible for the experiences associated with animal magnetism. Explaining talking tables proved to be more of a challenge, but they tended to be attributed to a particular state of animal magnetism called "artificial somnambulism." Puységur had discovered that in this state, patients could carry on a conversation and recover from a disease by talking about their experiences to their physician. He had claimed that subjects in a state of provoked somnambulism possessed heightened sensibilities: some had even predicted the future, described concealed objects, or communicated with the dead.[66] Magnetizers naturally associated mediums with patients in a state of artificial somnambulism, or somnambulists, and referred to an ethereal intelligence as responsible for the phenomena.[67] Louis Goupy, for example, argued that like every other mediumistic or somnambulistic wonder, the talking tables possessed an intelligible cause. He believed that each individual had what he described as a personal ether, which could be transferred in part to others. He theorized that mediums and somnambulists were particularly susceptible to energy transfers and thus drew on the personal ethers of those surrounding them, giving them a surplus of fluidic energy that allowed the production of various manifestations.[68]

By the end of the nineteenth century, occultists were also beginning to consider the concept of a human fluid to explain certain phenomena. If such a fluid existed, they postulated, it should be observable, even measurable, and various apparatuses were designed to do so. Writing in *L'initiation*, the occultist Horace Pelletier explored the possibility that a force similar to electricity was present in certain subjects—mediums, for example. Such a force would allow the displacement of objects without contact, including the deviation of magnetized needles. Pelletier experimented on this force in many situations and concluded that the intensity of the phenomena varied proportionally to the intensity of the force projected outside the body. The degree of the force could be affected, he argued, by the frame of mind of the subject at the time and the environment.[69] Paul Joire, president of the Société universelle d'études psychiques, talked of an exteriorized force that rendered subjects capable of moving distant objects. To observe and measure the force emanating from the nervous system, Joire invented the *sthénomètre*, an instrument consisting in a needle isolated under a glass globe. While experimenting with it, Joire observed that, in the case of a disease, the exteriorized force was proportional to the depression of the nervous system. More than simply measuring the capacity of an individual to produce mediumistic phenomena, the sthenometer offered medicine a great tool, he optimistically concluded.[70]

The physician Hippolyte Baraduc took a similar approach to Joire and created the *biomètre*, an instrument he believed would measure the movements or vibrations of the human soul. The biometer consisted of a nonmagnetic, isothermal needle placed on a board divided into three hundred and sixty degrees. When using it, Baraduc asked subjects to place their hands over the apparatus without touching it. He would then measure the displacement of the needle and the lapse of time during which the movement had been observed. The vibrations, he claimed, varied in degrees and duration depending on the subject's temperament and state of mind, as well as the exchange of fluids between this individual and the surrounding environment.[71] While the neurotic subject was disturbed by pathogenic vibrations, the sensitive one was able to perceive ethereal forces and use them in a healthy way. By providing a means to measure fluidic vibrations in individuals according to the degree and duration of the movement, the biometer supposedly allowed Baraduc to diagnose specific problems. He believed he had introduced quantification into this aspect of medicine. Now, mathematics and geometry could be used to diagnose nervous diseases. Just as a thermometer measured fever, he claimed, the biometer measured the human fluid; and, like a thermometer, it would assist the physician in assessing the severity of a disease.[72]

Believing that his biometer would explain miraculous healing, Baraduc sought signs of a fluid operating at Lourdes. He was convinced that the miracles happening at the sanctuary were authentic and could not be explained through the usual scientific concepts. Beyond electricity, human will, or drugs, there was, on occasion, the power of religious beliefs, which he explained as an exchange of fluids between a particular individual and his or her surroundings. At Lourdes, he postulated, the atmosphere created by an intense pilgrimage, the atmosphere of piety, and collective prayer allowed subjects to connect their individual selves with the sidereal potentialities of the universe. Curative effects were brought on by this connection.[73] With his biometer, Baraduc measured the external forces affecting the pilgrims, which he believed worked in ways similar to electricity. The large crowds in religious ecstasy clamoring "Hail Mary" in unison were responsible for the rise of a curative force that created a shock, reaching and curing patients in three possible ways: physically, physiologically, and psychologically.[74] Based on photographic plates he had placed in different locations around the sanctuary, Baraduc claimed to have found that the pools of Lourdes generated a force he called "salutary dew." It was this dew, produced by God, that caused the miraculous healings witnessed at the sanctuary.[75]

Others used photography in an attempt to record the existence of invisible entities. In 1912, two men from Bordeaux known only as Mesnard and Plomb developed a photographic apparatus supposedly able to record radiations emanating from unknown entities. They would dip a photographic plate into an unspecified liquid that supposedly rendered the plate more sensitive; then, in complete darkness, they would point a photographic apparatus toward a screen treated with radioactive liquids and produce photos in which faint discharges where made visible: "We think that we have made great advances in the domain of fluidic photography," they rejoiced, contending that photography was surely the impartial witness that would provide final and definite proof of the immortality of the soul.[76] Such work was not without its critics. The editors of *Annales des sciences psychiques* were very cautious about Mesnard and Plomb's claims as published in their own journal, pointing to a lack of details and a failure to provide a clear presentation of the procedure.[77] Jean Mondeil, an army captain, criticized both Baraduc and Mesnard's theories and argued that the instruments designed to measure the action of such a fluid were only measuring electricity. The peculiar photos obtained by occultists like Baraduc or Mesnard and Plomb were more likely the product of differences in temperature rather than proof of the existence of some special force.[78]

Guillaume Fontenay, a spiritist and professional photographer, also criticized the use of photography in attempts to observe a human fluid. Fontenay, who had also developed methods to detect fraud in spirit photography, showed that certain photographic phenomena were a product of the methods used in photography rather than supernatural occurrences. He argued that most of the photographic proofs obtained after prolonged exposure were caused by chemical reactions rather than occult forces. To prove his point, he demonstrated that the same effect could be produced by immersing a photographic plate and a piece of paper containing a few words in hot water or by having an individual apply the plate and the piece of paper to his forehead or his stomach. During development, the writing would appear on the photographic plate in both cases. The phenomenon was thus caused by ambient temperature and not by a mysterious human fluid. By studying the different types of papers and inks used to produce the effect, Fontenay was also able to account for the fact that the writing would sometimes develop in negative and sometimes in positive. Nothing about this particular phenomenon required more than a chemical explanation, Fontenay concluded, and he warned spiritists and occultists to use caution with photographic techniques that were not yet sufficiently well understood.[79]

One of the most distinguished and notable researchers to investigate human fluids and astral bodies was Lieutenant Colonel Albert de Rochas d'Aiglun, who after his retirement as a military engineer became an administrator of the École polytechnique in Paris, of which he was a graduate. Rochas developed an interest in the supernatural and attempted to prove that humans possessed a soul with an influence extending outside the physical body. This exteriorization of the sensitivity, as he called it, could be experienced as an aura or rays emanating out of the body. That such manifestations could be observed in some subjects was not a new claim. Visions and healings had often been associated with such sensibilities. In the Christian tradition, artists often represented the Virgin Mary with healing rays emanating from her fingers, and saints were painted with halos above their heads.[80] For Rochas, the time had come for this ancient faculty to be understood, quantified, and perhaps even controlled; and he believed that the key to this lay in hypnotism. Already, magnetizers had claimed that while in a trance, certain subjects experienced peculiar sensations or visions.[81] After performing his own experiments, Rochas agreed, and he concluded that subjects who experienced visions did so in one of the stages of hypnotism, which he called the state of connection. During this phase, several of them had momentarily become hyper-excited, allowing them to perceive auras or rays in others.[82]

Continuing his work, Rochas studied the ability in some to perceive electro-magnetic currents visually. On occasion, he found, subjects were able to see discharges and consistently describe colors as a current flowed in their proximity. In other experiments, Rochas worked on light refraction and the modification of auras by attempting to determine how the perception of a color changed when he placed his finger in a light source. Subjects reportedly observed a change in the colors perceived when he had put his fingers in the light without their knowledge. A series of experiments convinced him that finger radiations gave subjects a sensation of blue and violet. In yet another experiment, Rochas concluded that a powerful magnet could alter the coloration of the discharges reported by subjects. Also experimenting with polarization, he observed the ability in some to perceive changes in the polarity of a current.[83] Based on all of his experimental evidence, he concluded that external sensibilities were perceived through the eyes, that the nervous fluid producing such exteriorization of the sensibility was red or blue, and that it could be observed either as downy hairs covering the skin or as discharges escaping the various organs of the senses and the extremities of the body. Subjects described the discharges coming out of the body as flame, round if emanating from the whole body and long if emerging out of the extremities. The length, the intensity, and the coloration of the discharges was said to vary from subject to subject, and even in the same subject, depending on fatigue or the state of hypnotism, but were consistently observed.[84]

What Rochas called the externalization of the sensibility was the ability, reported in some hypnotized subjects, to experience heightened senses and memory. In particular, the sense of touch was said at times to be enhanced to the point where sensibility was no longer limited to the physical body. Hypnotists had observed a loss of sensation in hypnotized subjects, who could be pinched or made to smell ammonia without any effect. As smell and touch were affected, however, hearing and sight continued to function. After subjects had been under hypnosis for a certain time, they were said to regain their senses, but in an altered form. The sense of touch was no longer limited to the surface of the skin, but now spread to the surrounding environment. Memory was also affected, becoming specialized and selective, between periods of lethargy. This was an exteriorization of sensibilities.[85] Rochas believed that this phenomenon could perhaps be explained if hypnotism was understood to work like an electric current or an electromagnet. The current could cause subjects' nervous fluid to escape their bodies. This external fluid would then be distributed in maxima and minima in the form of a vibrating movement that might correspond to the rhythms of the heartbeat and breath.[86] He hoped that his concept of the exteriorization of the soul would provide a comprehensive basis for understanding mediumistic abilities and other phenomena of the kind.

∽

Like spiritists, occultists were an important part of the landscape of alternative spiritual movements in France at the time. Whether it was by appealing to astral bodies, fluids, or the exteriorization of the sensibilities, they looked for new ways to explain not only mediumship but a broad range of arcane phenomena. By incorporating ancient teachings into it, they hoped to push the concept of science into new territories. They attempted to develop a system of teachings based on revelations that would acquire an authority similar to that of scientific theories, a science of the magical and the unexplained, a new order of mystical symbols and meanings. Against the trends dominant in science at the time, occultists clung to the idea that there are things in this world that should be sacred and revered, phenomena that cannot be understood except by revelations, knowledge that should be kept secret, and wisdom that should be handled carefully.

Occultists claimed to possess the ultimate knowledge, but limited themselves to oral revelations. This esoteric characteristic of their movement makes any attempt at assessing their successes difficult. To the outside observer, French occultism does not seem to have been very coherent. Its doctrines appear confused, and its claims to secret knowledge overstated. Its members never agreed upon which ancient revelations to follow or how to interpret and adapt them to the modern world. In their claims to have privileged knowledge, they were at best vague and at worst insincere. In spite of this, occultists were an important part of the pervasive dissatisfaction with the positivist and materialistic attitudes that were felt to be dominating the scientific institutions of the period. But while occultists were appealing to astral bodies and ancient revelations to explain séances and other instances of the supernatural, an entirely different set of explanations, which treated everything spiritist and occult as pathological, was gaining popularity in hospitals and universities around the country.

CHAPTER THREE

Pathologies of the Supernatural

The city of Lyon was in danger, Dr. Philibert Burlet warned in a lecture at the Société des sciences médicales at the end of 1862 in which he reported on the cases of six patients at the Antiquaille hospital who he thought were exhibiting signs of mental illness directly related to the practice of spiritism. Moreover, this situation was not unique, Burlet told his audience—every physician in the region dealing with mental illness had already encountered similar cases. If this was true of the rest of France as well—and there was no reason to think otherwise—spiritism was well on its way to becoming one of the chief causes of mental alienation in the country.[1] While more religiously concerned thinkers had focused on the dangers that spiritism represented for the soul, physicians often diagnosed séances as pathological. Spiritism drove the mind to focus obsessively on certain thoughts and rendered the subject unable to function in normal, everyday life. From a medical perspective, Burlet stressed that the practice of spiritism had to be considered a mental illness, caused by the exaggeration of religious ideas, an intense belief in the supernatural, and an unhealthy love of the mysterious.[2]

Even as groups of spiritists were organizing themselves around the county, the human sciences were developing, professionalizing, and creating their own internal division of labor and specializations. Psychiatry and psychology were becoming increasingly influential in hospitals, universities, other establishments of research and higher education, and even the legal system. By the mid nineteenth century, psychiatry had become a recognized medical discipline. Psychology took somewhat longer to develop. Like psychiatrists, psychologists were interested in human experiences and behaviors, but, unlike psychiatrists, they were neither physicians nor necessarily affiliated with a hospital or an asylum. Rather, psychologists operated within the university system and the establishments of research and higher education. Officially, French psychology began with the creation of a chair of experimental and comparative psychology at the Collège de France in 1888. By the beginning of the twentieth century, it had become part of the academic landscape.[3]

Psychology differed from psychiatry in more than just its structural organization. Whereas psychiatrists focused on pathologies and cures, psychologists were interested in formulating explanations of human behaviors in more general terms. No matter what their approach, however, the two groups were never fully independent of each other. With the common aim of explaining human experiences and behaviors rationally, both disciplines were destined to intrude on a domain that had traditionally been the purview of the Church. Not only were physical manifestations of religiosity—possession, visions, cures, and stigmata, amongst others—of interest to them, but they were potentially problematic for sciences constructed on the assumption that the human mind and its productions could be explained physiologically. The success of psychiatrists and psychologists depended in part, not only on their abilities to account for these seemingly supernatural phenomena, experienced by a small portion of the population, but to have their explanations accepted in scientific circles.[4]

In their race to devise a master explanation of what was really happening at séances, medicine, psychiatry, and psychology each attempted to appropriate the supernatural and, in particular, mediumistic abilities. Each provided a way to accept the phenomena witnessed while rejecting the mystical interpretations usually assigned to them by subjects, audiences, priests, and spiritists. By presenting supernatural experiences as pathological in nature, medical doctors and psychologists legitimized them more than any other group, but they did so by reducing them to the physiological expressions of mental disorder. It was a disease of a few, but provided a window into the condition of all. Observing mediums and their ectoplasmic productions (spirit substances supposedly emerging out of a medium's body) would bring about a better understanding of the potential of the human mind in some of its most mystifying pathological manifestations. In particular, the psychologists Théodore Flournoy and Pierre Janet and the physician Joseph Grasset each formulated their own theories of the personality based on the experiences of mediums and reported cases of haunting. Then, between 1905 and 1908, the Institut général psychologique organized a series of séances with the Italian medium Eusapia Palladino, indicating an openness on the part of psychologists to consider mediumistic phenomena, as well as a willingness among mediums to be studied. On the whole, however, in their considerations of séances and other supernatural manifestations, scientists remained skeptical and often hostile. Even the most sympathetic of psychologists could not escape the patronizing attitude of their colleagues. In their eyes, mediums, stigmatics, and visionaries were patients, and spiritism joined possession, visions, and spiritual delusions as the latest expression of a dangerous delirium.

BETWEEN RELIGION AND PATHOLOGY

In Europe, the second half of the nineteenth century was a period filled with instances of the supernatural made concrete. Marian apparitions, miraculous cures, demonic possessions, stigmatics, and visionaries were frequently reported on in the press, as were mediums and somnambulists. Starting with Catherine Labouré's visions of the Virgin Mary in Paris in 1830 and those of Mélanie Mathieu and Maxim Giraud at La Salette in 1846, Marian apparitions increased in frequency throughout the century. In 1858, Bernadette Soubirous witnessed a series of apparitions in Lourdes that were among the most famous sightings of the period. Her visions led to the building of a shrine and launched an important pilgrimage tradition, coupled with frequent claims of miracle cures that continue to this day.[5]

These developments did not escape the notice of physicians, who showed a growing interest in such religious experiences. At La Salpêtrière hospital in Paris, the neurologist Jean-Martin Charcot adopted the old term "hysteria" and redefined it as a newly documented physiological condition associated with a multitude of behavioral symptoms, including religious delusions. This interest in religious manifestations on the part of the medical establishment was not new, but it was intensifying. In the early nineteenth century, the founders of psychiatry, Philippe Pinel and Jean-Étienne-Dominique Esquirol, had mentioned religious melancholy, but had predicted that it would disappear as science progressed. By the 1840s, however, the emergence of a more open and social Roman Catholicism began to threaten anticlerical medical officers, leading the latter, in turn, to develop a more sustained interest in demonology. In 1843, Maurice Macario thus described *démonomanie* as a dangerous inherited form of melancholia, or lypemania, spread by mental contagion or imitation, and more common outside of Paris than in the capital. This view continued to gain popularity in medicine, and, by the 1880s, the group around Charcot had become convinced of the pathological character of any form of religious devotion. In an attempt to widen the scope of their research, physicians at La Salpêtrière began to explain not only contemporary but also past cases of possession, stigmata, and other tangible manifestations of faith by invoking Charcot's characterization of hysteria.

The practice of reinterpreting past religious occurrences in scientific terms had first been launched by the ardent positivist Émile Littré, best remembered today for his *Dictionnaire de la langue française* (1863–73), in an 1869 article on early modern accounts of miraculous cures associated with the bones of Saint Louis (King Louis XI of France).[6] Retrospective medicine, as he called to it, was later taken up by one of Charcot's disciples, Désiré-Magloire Bourneville. Between 1882 and 1902, Bourneville published a series of ten volumes, collectively titled La bibliothèque diabolique, providing various reinterpretations of classic cases of witchcraft, demonic possession, and other mystical experiences. Noted for his attacks on mysticism and supernatural beliefs, Bourneville would also become famous for his anticlerical stance and his defense of both cremation and the laicization of nursing.[7] The Bibliothèque diabolique consisted of republished accounts of past events, as well as contemporary reports of religious manifestations, each explained in terms of hysteria. In most cases, prefaces and commentaries were added to centuries-old manuscripts. In them, Bourneville and other physicians rejected the earlier claims of possession in favor of pathological explanations of *hystéro-démonopathie* or *hystéro-épilepsie*. In *Procès-verbal fait pour délivrer une fille possédée par le malin esprit à Louviers*, an old manuscript from the Bibliothèque nationale, edited by Armand Bénet and supplied with a new introduction, a case of possession and exorcism was reinterpreted in clinical terms: "The supernatural in pathology and therapeutics is a myth, or, rather, a scientific heresy," it was proclaimed. "The facts are real, the conclusion is false: nonexistent extranatural intervertion—necessary at the time as an explanatory hypothesis—was deduced from the physiological and pathological phenomena."[8]

By the 1890s, the tone had changed slightly. Charcot himself wrote an article on faith healing, "La foi qui guérit," originally published in both English and French. It allowed for the possibility of faith healing or miracle cures that eluded medical or scientific explanation. For Charcot, however, such miracles were always hysterical in origin. They developed during a long, intense pilgrimage to a sanctuary, filled with powerful, suggestive thoughts on healing. As such, faith healing might work on certain diseases, but only in highly suggestible subjects.[9] Pierre Janet, a student of Charcot's, agreed with his teacher that many of those who made claims to tangible religious experiences such as miracle cures or stigmata were in fact suffering from nervous conditions. For physicians, the thoughts and experiences of such patients could be useful in revealing hidden pathologies of the mind. In the last years of the nineteenth century, Janet developed a close working relation with one such patient: Madeleine, a devout Christian who exhibited a profound sense of mysticism, a deep devotion to Christ, and occasional stigmata. In *De l'angoisse à l'extase*, published in two volumes in the 1920s, Janet related his encounters with Madeleine during her stay at La Salpêtrière and beyond. He described her case as psychopathological in nature, but recognized and respected her profound religious faith. By then, psychiatry had moved a long way from Bourneville and his Bibliothèque diabolique's portrayal of all mystics as deeply disturbed individuals under the power of their outdated convictions.[10]

Not all physicians rejected the potential power of fervent faith and the possibility of miracles as easily as those at La Salpêtrière did. Some even accepted them readily. Often holding regional positions, and in contrast with the rampant anticlericalism prevailing in Paris, these physicians were devout Catholics and often monarchists hoping for the restoration of the crown and the authority of the Church. As such, visions and prophesies proclaiming the eminent return of the monarchy and the ascent of the legitimist pretender to the French throne were of particular interest to those natural scientists in search of tangible evidence of their faith.[11] The 1880s and 1890s saw an effort on their part to construct a Catholic science to oppose the secular medicine developing in Paris, particularly around Charcot. In 1888, the first of five Congrès scientifiques des catholiques sanctioned by Pope Leo XIII was held in an attempt to answer the biological and medical sciences. At the same time, Church officials began to ask the help of scientists with some of the more extreme cases of mystical phenomena. In the 1860s, for example, L.-J.-J. Constans, *inspecteur-général* of the French national asylums, and the alienist L.-F. Calmeil were sent to the small town of Morzine in the Alps to deal with an asserted outbreak of demonic possession; and in 1883, a medical consultation office was instituted at Lourdes, making the collaboration of physicians in certifying miracles there official.[12]

Catholic physicians did not dismiss the possibility of miracles and wrestled with ways to understand the teachings of the Church in light of the latest scientific theories. The physician Félix de Backer wrote that his visit to Lourdes had convinced him that healings could occur outside of science. Such miracles had to be accepted by physicians.[13] In his study on stigmatization, the physician and professor at the preparatory school of medicine of Clermont-Ferrand Antoine Imbert-Gourbeyre, struggled to reconcile his faith and his belief in miracles with his scientific training. If the Church declared a phenomenon to be miraculous, he argued, scientists had to accept the verdict and renounce their own authority on these matters; whereas false cases of stigmata could be explained by science, miracles could not. The power of imagination could not act on blood in such a remarkable way; it could never account for the appearance of wounds at specific and religiously significant sites on the body.[14]

In recognizing the supremacy of the Church on matters of faith, Imbert-Goubeyre illustrates the conflict experienced by Catholic physicians.[15] For some of them, faith ruled over science. Alexandre Jenniard du Dot, for example, believed that only theologians had the right to explain spiritism. He also warned of the dangers of suggestion, as some physicians practiced it, because it substituted the subject's will for that of the physician: "it is *doctoral possession* substituted for the old *diabolical possession*."[16] The physician Charles Hélot, medical expert for the Rouen diocese in cases of possession since the 1870s, also cautioned fellow scientists against the dangers of natural theories invented by an atheistic science hoping to negate the will of God and all manifestations of the supernatural. He argued that the phenomena of somnambulism, hypnotism, and haunting should not be discussed in scientific terms when religion could sufficiently account for the manifestations. Like du Dot, Hélot believed that hypnotism and suggestion were dangerous. They constituted the implicit invocation of the devil brought about by a physician with a desire to provoke a state only the devil would wish to induce.[17]

Aside from Catholic mysticism, physicians and psychiatrists were also interested in séances and their mediums. Whereas the more faith-based physicians were placed in an awkward and contradictory position regarding such manifestations, the vast majority of them saw nothing more than physiology and dangerous pathology in such phenomena. In 1904, Paul Duhem warned: "Spiritism must not be seen as a simple and innocent societal game but rather as a danger that we, as physicians, should attempt to destroy."[18] By then, spiritism had already been associated with mental illness for a few decades. Physicians like Duhem and Burlet argued that to practice spiritism, to take part in séances, would put the more susceptible subjects at great risk. If, for most participants, spiritism did no more than console and dazzle, it could have devastating effects on the sanity of more impressionable observers.

Where Catholic thinkers like Jules-Eudes Mirville, André Pezzani, and Henri Carion had stressed the perils of spiritism for the soul, many physicians stressed the potential dangers to the mind and described the types of deliriums that could be caused by such a practice. Two types of pathologies tended to be associated with spiritism: first, for those subjects already predisposed to crisis, spiritism could provoke a delirious episode that would not necessarily have arisen under different circumstances; and second, for the more unstable subjects, spiritism provided the outlet for a madness to which they had been predestined.[19] Some patients experienced crisis brought on by their newly discovered mediumistic abilities. Others incorporated elements of the spiritist doctrine into their delirium. For physicians, the frequent practice of spiritism could lead to a doubling or even what they called a *déségrégation* (disaggregation) of the personality, a pathological state in which the conscious and the subconscious separated themselves to evolve separately. This was often described as the greatest danger for the participants at a séance. In cases of complete *déségrégation*, unconscious acts would appear to be conscious, but would take place without the awareness of the person performing them. For many physicians, this became the key explanation of spiritist productions.[20] Mediumship was thus a disease of the mind. The phenomena observed at séances were sometimes real, but they were never supernatural; they were the product of patients, not of spirits.

Not all participants at séances would experience such florid symptom of mental illness, of course. Healthy ones would be relatively safe. But it was argued that spiritists had a moral obligation to refuse initiation to those who were susceptible to pathological crises. If spiritism wanted to be seen as a serious practice, it would have to limit itself to sane, skeptical subjects who came to learn or for amusement and prevent those who turned séances into delirious, hallucinatory circuses from attending.[21] Of course, such warnings angered spiritists, who conceded that, just like anywhere else, there were mentally ill people in their ranks, but denied that spiritism caused mental illness. If they had any responsibility, it was to soothe perhaps already diseased imaginations and nothing more.[22]

THÉODORE FLOURNOY AND HÉLÈNE SMITH

In 1899, the internationally known Swiss psychologist Théodore Flournoy (1854–1920) presented the results of his five years of observation of the medium Hélène Smith in his book *Des Indes à la planète Mars: Étude sur un cas de somnambulisme avec glossolalie*, one of the most important studies of mediumship ever published. Flournoy had been a student of experimental psychology under Wilhelm Wundt in Leipzig before obtaining a professorship in psychophysiology at the University of Geneva in 1891. Throughout his career, he wrote numerous books and articles on religious psychology and came to be considered an expert on the topic.

Des Indes à la planète Mars sold well both in French and in the English translation, which came out in 1900.[23] "Upon my word, dear Flournoy, you have done a bigger thing here than you know; and I think that your volume has probably made the decisive step in converting psychical research into a respectable science," his friend the American psychologist William James wrote to him that year.[24] Flournoy had provided a scientific account of séances that, while remaining respectful of Smith and her beliefs, had presented her in clear pathological terms. At the turn of the century, Des Indes à la planète Mars became a landmark for mediumistic theories outside spiritist and occultist circles.[25]

Flournoy was neither the first psychologist to develop an interest in mediumship nor the only one to discuss the phenomena from a physiological perspective. By the time Des Indes à la planète Mars came out, participation at séances had already been considered a trigger for mental illness for decades. In his focus on a single medium and his efforts to develop a caring relationship with her, however, Flournoy differed from most of his colleagues. In this, he had more in common with psychical researchers who treated their subjects with respect and admiration than with those who treated them as patients.[26] If he did not share Smith's views on the cause of her visions, he never doubted her sincerity. "It is clear that I would not have considered such an enterprise with just anybody," he confided to James:

> For one thing, there could be no question of surrendering my freedom to think and write in accordance with my ideas; but how many mediums would agree to see their phenomena exposed and explained in a more or less scientific way, that is to say, in a very different manner than the one prevailing in the spiritist circles in which their abilities are developed? In this particular case, fortunately, the difficulty appeared less significant thanks to the strong and distinguished character of the medium with whom I was dealing. Miss Smith appeared in fact to be a remarkably intelligent and gifted person, much above ordinary prejudices, very open and independent of thought, and consequently capable of accepting, simply for the love of truth and progress in research, that we turn her mediumship into a psychological study, at the risk of obtaining results that are not in keeping with her personal impressions and the opinion of her circle.[27]

By the time Flournoy met Smith in 1894, he had already been trying to gain access to spiritist circles in Geneva for some time. Intrigued by mediumistic phenomena, he had hoped to observe a séance and verify the spiritists' claims for himself. "I try to penetrate into the spiritualistic world of our city, but it is rather difficult," he wrote James in December 1893.[28] A few months later, he met Hélène Smith (the pseudonym of Elise Catherine Müller), a worker in a silk shop in Geneva. In September 1895, writing to James once again, Flournoy recounted his first impressions:

I was forgetting to tell you what has interested me most during the last six months: it is a certain medium (nonprofessional, unpaid) of a spiritualist group, into which they have agreed to accept me in spite of my neutral position; I have attended about twenty of the séances of which a third were here at my home; psychologically, it is very interesting, because this woman is a veritable museum of all possible phenomena and has a repertoire of illimitable variety: she makes the table talk,—she hears voices,—she has visions, hallucinations, tactile and olfactory,—automatic writing—sometimes complete somnambulism, catalepsy, trances, etc.[29]

At their first séance together, Smith impressed the psychologist by revealing her knowledge of deceased members of his family. As they were both from the same city, Flournoy suspected these must have come from conversations overheard by Smith in her childhood and unconsciously remembered: "The great majority of the phenomena were evidently the automatic reproduction of forgotten memories—or memories registered unconsciously. There is actually in the nature of this medium a second personality who perceives and recall instances which escape ordinary awareness," he wrote to James. "What is irritating in this kind of observation is the difficulty of making it precise, the medium and the members of the group having a holy terror of everything which resembles an 'experiment.'"[30] Fortunately, the problem soon disappeared, because Flournoy was able to persuade Smith to perform for him privately.

For five years, Flournoy investigated Smith's trance-state visions. Unlike other scientists interested in mediums, he did not focus on physical phenomena or communication with another realm. Instead, he decided to follow the traces of Smith's unconscious as they leaked out during trances in which she revealed information about two of her supposed previous lives. Smith believed herself to have been the daughter of an Arab sheik who under the name of Simandini became the favorite wife of a Hindu prince, Sivrouka Nayaka, who had reigned over a region called Kanara and built a fortress called Tchandraguiri in the year 1401. Later, she was reincarnated as Marie-Antoinette, and now, as a punishment for her past sins and to perfect her character, she had come back in the humble form of Hélène Smith.[31] While in a state of somnambulism, Smith also revealed the existence of her spirit guide, Léopold, who was none other than the illustrious Count Alessandro di Cagliostro, who, after a long search, had again found the object of his affections in the reincarnation of his dear Marie-Antoinette. Flournoy described Léopold as a dominating spirit guide, witnessing and controlling most of Smith's trances.[32]

A sample of Hélène Smith's writings in Martian. From Théodore Flournoy, *Des Indes à la planète Mars. Étude sur un cas de somnambulisme avec glossolalie* (Paris: Le Seuil, 1900), fig. 27.

Smith had only recently been initiated into spiritism, but, although her mediumistic abilities officially dated from 1892, Flournoy believed that she had been showing signs of her potential since childhood. Smith herself claimed to have experienced her first visions at the age of ten, when Léopold had originally revealed himself to her. It was only once she discovered spiritism, however, that she came to understand fully these previous experiences. Based on her recollections,

Flournoy suggested that during puberty, Smith had developed a tendency toward mental *déségrégation*, from which she had progressively recovered. In fact, if she had never been initiated to spiritism, he maintained, her personality would have gradually reconsolidated, and with time her visions would have disappeared. In spiritism, unfortunately, Smith's tendencies had found a fertile space for expression. He feared that her participation at séances had pushed her toward increased mental *déségrégation*. Eventually, it had led her personality to split between an awake and a trancelike self.[33]

Aside from past existences, Smith's visions had also taken her out into the solar system, where she witnessed life on the planet Mars: "Mlle Smith, by virtue of the mediumistic faculties, which are the appendage and the consolation of her present life, has been able to enter into relation with the people and affairs of the planet Mars, and unveil their mysteries to us," Flournoy wrote.[34] He devoted most of his book to what he called the "Martian romance" and attempted to show such a story to be the product of his medium's imagination. Smith had described life on Mars in detail, providing him with drawings of the landscapes and cities, and even speaking and writing in what she claimed was the Martian language. Glossolalia, the ability to speak and understand a language that one has not been taught, has been claimed by mystics since the early days of the Church. In the case of Smith's Martian, however, Flournoy was confronted with an unknown language, whose phonetics, grammar, syntax, vocabulary, and style he studied: "It's a typical case of 'glosso-poetry,' of complete fabrication of all the parts of a new language by a subconscious activity," he concluded.[35] He described Smith's Martian as a natural language, one that had been created unconsciously by her second personality.[36] It turned out to be very similar to French: each Martian letter had its equivalent in the Latin alphabet; Martian was composed of articulate sounds, all of which, consonants as well as vowels, existed in French. Moreover, word order in Martian was identical to French.[37]

Smith's construction of a new language and her seeming knowledge of Sanskrit even attracted the attention of linguists.[38] Already at the time of his observations, Flournoy had asked the eminent linguist Ferdinand de Saussure to shed some light on the curious phenomena. In 1901, intrigued by Smith's claims, the linguist Victor Henry also worked on her Martian and published his analysis of the language. The similarities between French and Martian were not always obvious, he wrote. The different vocabularies made them hard to discern at times, but they were undoubtedly there. In fact, Henry claimed that the two languages were completely grammatically identical, down to the use of the auxiliary verbs "to have" and "to be."[39]

The publication of *Des Indes à la planète Mars* in 1899 created some contro-versy. The spiritists of Geneva were not very happy with Flournoy's presentation of his medium, nor were they pleased with their newly found notoriety across Europe and America.[40] They quickly published a reply to Flournoy's study, *Autour «Des Indes à la planète Mars»*, in which they accused him of failing to see the authenticity of Smith's production. A series of heated exchanges in which each side attacked and ridiculed the other followed.[41] As for Smith, although Flournoy had thanked her in his book for her openness to his account of her cycles, she now understood the extent to which she had allowed him to interpret her trances. Their relationship began to turn sour when, soon after the publication, a dispute over the revenues of the book exploded. According to Olivier Flournoy, grandson of the author, Flournoy had promised to give all of the profits to Smith. When her new fame led an enthusiastic widow to donate some money to her, allowing her to live a comfortable life, however, Flournoy decided to split the earnings be-tween his medium and the newly founded *Archives de psychologie*, which he was then running with his cousin, Édouard Claparède.[42] The question of the reve-nues caused a ten-year feud between Smith and Flournoy. By 1909, tired of all of it, the psychologist wrote Smith one last letter in which he ended the relationship definitely by giving her all the profits from the fourth edition of *Des Indes à la planète Mars*.[43] He appears to have had no further contact with Smith.

Flournoy's final letter to her seems to have caused Smith to have a nervous breakdown. In a séance with her psychiatrist at the time, the sexual undertone once implied by her spirit guide Léopold now became overt. In a letter to Flour-noy, the psychiatrist who witnessed Smith's deterioration recalled one session in particular: "getting more and more excited during her story, the medium takes a lascivious attitude; the eyes are languishing, the bust tilted back, the hands *active* and finally . . . H. S. is taken by an erotic spasm that leaves no doubt on the illu-sion of a sexual connection."[44] Smith's sexual confessions would continue, be-coming even more explicit four days later, her psychiatrist reported. These final sessions would mark a turn in the medium's life. Afterward, she distanced herself from spiritists, claiming she found them too dominating. She began a second mediumistic life, as it came to be called, devoting herself to painting the mysteri-ous worlds she had once so vividly experienced in séances.[45] She died in 1929, isolated in her own world of mysticism.

As for Flournoy, he continued to pursue his interest in mediumistic abilities even after his break with Smith. In the following years, he attempted to generalize the conclusions he had reached with her. During séances, he concluded, medi-ums entered a psycho-physiological state that encouraged mental dissociation and a regression to an inferior state ruled by their imaginations. Many of the phenomena witnessed could be explained as latent memories, instinctive tendencies, and other resources of the subconscious. Mediumistic phenomena had nothing to do with the dead, but in fact consisted in a set of mental processes and "still-mysterious laws."[46] Flournoy recognized that the religious sentiment fulfilled a need in many people. Mostly harmless and destined to disappear and be replaced by rational explanations, it could, for the time being, turn pathological in some cases. "Religious life has everything to gain by no longer being confused with these orchestral manifestations that come with it in the more or less pathological temperaments," he wrote.[47]

Although *Des Indes à la planète Mars* aroused considerable interest at the turn of the century, it was rapidly forgotten. Flournoy's journey into the unconscious of Hélène Smith was ultimately eclipsed by the theories of his contemporary Sigmund Freud, whose work went beyond theorizing about the unconscious to offer a potential cure for psychopathological disorders.[48] Moreover, Flournoy's work was difficult to reproduce. In Smith, he had found a rare medium, one that was willing to be observed over a prolonged period of time and whose psychical manifestations were intellectual rather than physical. Others proposed studies on mediumship, but took a very different approach, focusing on the more spectacular physical phenomena produced. Among them, Pierre Janet and Joseph Grasset both considered the productions of mediums in formulating more complete theories of personality.

PIERRE JANET, JOSEPH GRASSET, AND THE MEDIUMISTIC PERSONALITY

In 1889, Pierre Janet wrote his doctoral thesis based on observations of female patients suffering from nervous pathologies, mostly hysteria. Janet was interested in automatisms that manifested themselves in catalepsy, somnambulism, suggestion, alternating memories, unconscious acts, and psychological *déségrégation*. In *L'automatisme psychologique*, he attributed mediumship to the *déségrégation* of the mind, whereby a set of thoughts formed outside of perception and unrelated to it. As such, mediumistic phenomena were produced by unconscious involuntary actions.[49] Janet saw striking similarities between mediums and patients exhibiting double personalities. Some forty years earlier, Eugène Azam, a professor at the medical faculty of Bordeaux, had first observed a case of double or alternate personalities. Since the age of sixteen, his patient Félida had been experiencing strange episodes of a kind of somnambulism in which her personality was altered;

she was happier and more carefree, even agreeing to sexual relations that had led to an out-of-wedlock pregnancy. In 1876, Azam had published his observations of Félida.[50] Janet's subsequent views on mediumship owed a great deal to this work. He argued that the most successful mediums exhibited clear signs of such psychological *déségrégation*, behaving as if they had two separate and independent personalities. More than simple curiosities, mediums were, for Janet, evidence of extreme importance in the study of the human mind.[51]

Inspired by this work, Joseph Grasset, a professor at the reputable medical clinic of the Université de Montpellier, developed a theory of personality based on the concept of *déségrégation* using his own observations of mediums. In 1903, the *Annales des sciences psychiques* published his article "Le spiritisme devant la science," in which he discussed the case of Jeanne, a fifteen-year-old girl around whom phenomena associated with a haunting had already been occurring for some time. He described the teenager as a hysteric, and the phenomena as the manifestations of her pathology. For Grasset, Jeanne's condition belonged to a group of phenomena yet to be understood by science. He believed, however, that mental suggestion, clairvoyance, telepathy, and, in Jeanne's cases, moving objects without contact with them (called telekinesis today) would soon be provided with a scientific explanation.[52]

By the time Grasset became interested in Jeanne, news of her feats had already stirred some interest. The hairdresser in her village had even written to the occultist magazine *Le messager de l'occulte* informing the readers of the peculiar occurrences:

> For eighty days now, there have been very extraordinary things happening in this house; as soon as these people heads are turned, the blankets, the sheets, and the mattresses are thrown into the center of the room, the chairs and tables overturned, the blankets carried out to the middle of the yard. . . . The dog, which was locked up, was found outside without anyone having opened the door; three days ago, the 15-year-old girl had her hair cut while she was in bed. The chrysanthemums, lilies, gillyflowers, and shallots have been devastated. The children, aged fifteen, six, and four, attest to having seen plants being destroyed with no one around; they have also seen a closet open up and clothes fall at their feet; at night, the walls and the furniture produce knocks.[53]

Commenting on the case, the editor of *Le messager de l'occulte* replied that, like this one, most cases of haunting took place in houses inhabited by young teenage females unconsciously acting as mediums and provoking these phenomena. To stop the haunting, he recommended that an iron sword be used to pierce the air around the house. This would destroy the electric clouds that had been created by the teenage medium and were responsible for the various phenomena. Grasset did not reveal whether or not the method was used in this particular case, but it was certainly a contemporary remedy for hauntings.[54]

For Grasset, hauntings were neither manifestations of the dead—or of Kardec's *périsprit*—nor invariably deceitful. In fact, they could often be explained through available concepts of physiology. Observing Jeanne, the physician was able to develop a theory of two psychisms associated with two different modes of intellectual activity, superior and inferior. The superior center constituted "the true and complete conscience" from which free will emanated. It was the center of personality, intellectuality, and the superior psyche. The inferior center was a polygon formed by six centers, each responsible for a specific activity: audition, vision, general sensitivity, movement, speech, and writing.[55] It was the site of automatism, inferior psyche, and sensations.[56] With this holistic theory, Grasset elaborated a controversial assertion that explained human behavior in its entirety. For Grasset, human behavior was governed by a superior center, responsible for superior thinking, and a polygon, the site of lower thinking, from which instincts, acts of passion, and acts of habit all emanated. For each action, a corresponding set of centers, or neurons, would every time be involved.[57]

Grasset's theory could potentially provide an explanation for the paralysis caused by organic lesions. He classified the various types of aphasias as lesions situated in different sites in his system: in the polygonal centers of speech, between the polygon and the superior center, on the fibers that united the various polygonal centers, or below the polygon.[58] Both the superior center and the polygon were physiological systems, which functioned together in the healthy body. However, they could temporarily or permanently separate. This separation of the superior center from the inferior polygon did not imply a complete cessation of psychical activity but only of that of a higher level. As consciousness resided in the superior center, awareness always depended on whether or not it was active. Acts involving only the polygon were labeled automatism. They were always spontaneous and never voluntary.[59]

Déségrégation of the personality could be partially experienced during sleep or in moments of distraction, during which the polygon took over behavior but not all functions. Whereas in such situations, the superior center continued to work properly, certain pathologies were characterized by its deterioration. In such instances, it became sick and ceased to function adequately. Only in such states— hysteria, somnambulism, and ambulatory automatism, for example—did the separation of the superior center from the polygon become observable.[60] In the states of hypnotism and suggestibility, the subject's superior center completely separated from his or her polygon, to be replaced by the magnetizer's own superior center. Thus, contrary to persuasion, in which one's superior center was being convinced by someone else, in suggestion, the subject's superior center became disconnected, inasmuch as consciousness was taken over by the magnetizer.[61]

Aside from explaining sleep, dreams, hypnotism, and suggestibility, Grasset aimed to provide explanations for occurrences of the supernatural in general and spiritist phenomena in particular. Cases of moving tables were easily explained by Grasset in terms of unconscious and involuntary movements, because the polygon was put in a state in which it provoked movement without being given a voluntary order from the superior center.[62] The addition of the production of each of those small involuntary movements would, when superimposed, produce the significant effect by which the table would move.[63] The theory could also be used to account for phenomena associated with the pendulum and the divining rod: through intense concentration, the subject was able to direct his or her thoughts to the action and execute the movement without the knowledge of the superior center. The same mechanism could explain contact mind reading.[64] In the case of mediums, the phenomena were produced through a combination of the separation of the two centers and increased polygonal activity. The trance the medium entered consisted in the momentary splitting of the polygon from the superior center, or a *déségrégation* of the personality, and an increase in the activities of the six centers of the polygon. The greater the polygonal activity; the greater the medium.[65] Mediumship was thus a form of provoked and temporary neurosis, similar to spontaneous or provoked somnambulism. Although the superior center remained constant, the medium's polygonal personalities would vary according to momentary inspiration and suggestion, external or internal. For Grasset this constituted a clear pathology.[66]

Of course, spiritists themselves did not associate mediumship with pathology. In fact, they often defended themselves against such attacks. Kardec had addressed his own concerns with this in his *Livre des médiums*. Mediumship required energy and could cause fatigue, but the practice of spiritism did not lead to madness. If there was madness, however, spiritism could provide a context in which to express it. Thus, children, idiots, and anyone experiencing symptoms of eccentricity or a predisposition to insanity should be discouraged from the practice.[67] Some physicians, more sympathetic to spiritism, refused to discuss mediums in the context of pathology. The physician and psychical researcher Gustave Geley warned against confusing abnormality with pathology. For him, a serious study of mediumship would contribute to the development of an abnormal psychology. Hysterics and mediums were very different, but not sick.[68] Geley's criticism of Grasset's theory was that it provided only half of an explanation and failed to account for all of spiritism. While Geley did appreciate the strength of theories of automatism in explaining certain phenomena—the moving tables and the divining rod among others—he did not believe they could provide an explanation for the most complex phenomena of telepathy, suggestion, clairvoyance, lucidity, exteriorization, and materialization, to name a few. Moreover, Geley felt that Grasset had failed to provide a satisfactory explanation for the separation of the polygon from its superior center.[69] It was his contention that such a problem could disappear if one conceived of the polygon and its centers as associated with different parts of the body and separable. The polygon could be related to the physiological brain and the superior center to the superior principal of the being, independent of the organism. In this way, the superior center could have an action outside the senses, muscles, and brain, which could possibly explain higher mediumship adequately.[70]

Flournoy, Janet, Grasset, and Geley were not the only physicians and psychologists writing on the leveling of the different activities of the mind. In Britain, the psychical researcher and poet Frederic Myers sought to explain mediumship by appealing to a theory of the subliminal self with multiple levels of selfhood. For Myers, such multiple selves could lead an individual to great creativity. If mediumship did not inspire a plethora of works in the human and the medical sciences, however, the few works available on the subject were well received. Most notably, the work of the Institut général psychologique in Paris emerged as one of the most elaborate attempts to legitimize mediumistic phenomena within the field of psychology.

EUSAPIA PALLADINO AT THE
INSTITUT GÉNÉRAL PSYCHOLOGIQUE

The idea of creating an institute in Paris dedicated to using a psychological approach to the study of psychical phenomena was first mentioned in the early months of 1900. On June 30 of that year, the Institut psychique international held its first meeting. Twenty-two members were present, among them the philosopher Henri Bergson; the physiologists Charles Richet and Étienne-Jules Marey; the physicist Louis-Paul Cailletet; the chemist Émile Duclaux; the editor of *Annales des sciences psychiques*, Xavier Dariex; the president of the Société universelle d'études psychiques, Paul Joire; and the canon of Notre Dame, Ferdinand Brettes. The participants all accepted the program set forth for the new institute by the psychologist Pierre Janet. Unlike the physical sciences, which had been progressing for centuries, the sciences of the mind were fairly young, Janet observed, but dealing as they did with the human psyche and the connections between the corporeal body and morality, they were even more important than the physical sciences. They would one day contribute to the fields of criminology, education, and pathology by providing us with an understanding of social relations and human behavior. More than this, the sciences of the mind were key to the most fundamental questions about our nature.[71] Telepathy, telekinesis, lucidity, split consciousnesses, suggestion, and mediumship, as well as other, similar phenomena appeared to be associated with the deeper powers of the mind. Their study would bring scientists closer to an understanding of human nature. The Institut psychique international sought to help bring this about.[72] It would be devoted to the study of psychical phenomena without any preconceptions. All schools of thought on these manifestations would be admitted.[73] Above all, men of impeccable scientific credentials would finally study animal magnetism, telepathy, lucidity, and mediumship using proper resources and experimental methods.[74]

The newly formed institute was discussed at the Quatrième congrès international de psychologie, held in Paris in August 1900, in the two sessions at the congress that had been designated to address psychical research at least indirectly ("Studies Relative to the Phenomena of Somnambulism" and "Psychology of Hypnotism, Suggestion, and Closely Related Questions.") Even spiritists and occultists were granted a voice in these sessions, inasmuch as Gabriel Delanne, Léon Denis, Papus, and Hippolyte Baraduc all participated in the proceedings. The reception of their presentations by other participants, however, remained cold, even hostile at times. The psychologist Nicolas Vaschide declared that Delanne and Denis's presentations lacked any trace of scientific research. "These are literary impressions, confessions, and some professions of faith, entangled with a regrettable ignorance of scientific documents recorded by psychology in the past few years," Vaschide said.[75] On such studies in general, the German psychiatrist Oskar Vogt declared: "I protest first in the name of science and of psychology in general. I protest then especially in the name of suggestion and hypnotism. No sooner have we succeeded in having the reality of suggestion and hypnotism recognized, no sooner have we succeeded in launching, starting with these phenomena, a psycho-pathogeny, a psycho-therapy, and a psycho-hygiene in a greater sense, than the spiritists invade our session and compromise it with anti-scientific communications."[76] These outbursts were followed by other comments of disapproval regarding the place of such presentations at the congress.

Whereas the participation of spiritists and occultists at the congress was not well received, news of the new Institut psychique international fared much better. The psychologist and spiritist sympathizer Julien Ochorowicz discussed the mandate of the institute and proposed a less controversial name for it. The Institut psychologique, he suggested, would be a more appropriate name for a permanent international center for all psychological research, whether accepted or as yet unverified.[77] Questions on human nature, Ochorowicz explained, had been brought back to the forefront of scientific research with the development of hypnotism and the appearance of a new category of peculiar phenomena. The new institute was meant to unify individual efforts in the study of these new phenomena.[78]

Discussion at the congress revealed a receptive audience in psychologists. In a presentation of his observations on spiritist phenomena, Flournoy described his enthusiasm for an institute devoted to the study of psychical phenomena:

> Far from fearing that the institute will be preoccupied with spiritism and occultism, I believe that it is exactly this domain, loved by some and despised by others, that should be the archetypal object of these impartial investigations and the main aim of all these efforts. As I understand it, the idea of the founders of the institute is to introduce rigorous experimental methods into the study of these allegedly *supranormal* (to sum them up with one word) phenomena, and eventually to shine on them the full light of a science that is as yet hopelessly chaotic and murky; and it is to a project understood in these terms that I have given my support, in the conviction that such an enterprise answers a general and pressing need of our time.[79]

Officially founded on March 29, 1901, the newly renamed Institut général psychologique (IGP) now defined its purpose in a very broad manner: "to bring together in a monthly assembly the men of science who are occupied with the mind, its conditions, its own laws, its diseases, and its history."[80] Although no longer dedicated solely to the study of psychical phenomena, the members of the IGP would continue to include them as an important part of their broader interests. In its December meeting of that year, it was decided that study groups should be formed to address the different interests of the members. Four groups were initially created: collective psychology, moral and criminal psychology, psychical phenomena, and zoological psychology. Members of the group for the study of psychical phenomena were Arsène d'Arsonval, Henri Bergson, Édouard Branly, François Brissaud, Émile Duclaux, Étienne-Jules Marey, and Georges Weiss, all members of reputable academies and holders of respected academic positions.

Their section of the IGP was to be dedicated to the study of the unknown and, more specifically, to the exploration of the "still undefined forces" at the frontiers of psychology, biology, and physics.[81]

The first meeting of the section on psychical phenomena was held at Duclaux's home in January 1902, when it was decided to set up a laboratory to pursue observational and experimental studies. An appeal was made for all to report any psychical phenomena and the men or women who could produce them to the group. But, in its first few years of existence, very little was made of the section, something that dampened the spirits of psychical researchers.[82] News that the IGP had been rapidly transformed from an institute dedicated to the study of psychical phenomena alone to an institute of psychology had not gone unnoticed in psychical research circles. The creation of the IGP had been received with enthusiasm. Many had even contributed financially to it and were angered by the change of direction of the institute.[83] By 1902, psychical researchers had bitterly come to accept the nature of the new institute. The *Revue des études psychiques* reported: "So this is where we are. A *psychical* Institute has been created, mostly to study *psychical* phenomena; for this purpose, millions and millions of francs have been gathered and have been used to create a magazine open to all branches of psychology, but where psychical phenomena will be mentioned only when Monsieur P. Janet sees fit to ridicule them."[84]

Just a few years later, however, the IGP's section dedicated to the study of psychical phenomena began its most famous project: a three-year experiment with a single medium. In a meeting of the institute in March 1906, d'Arsonval, president of the institute and the section dedicated to psychical phenomena, for the first time officially discussed the experiments being performed by the section with the Italian medium Eusapia Palladino. The results were encouraging, he declared. At the same meeting, d'Arsonval also announced that the financial situation of the IGP was beginning to improve. The French government had agreed to grant the institute four million francs to build well-equipped laboratories, a library, and a museum.[85]

The séances with Eusapia were to be the institute's most ambitious project. From 1905 to 1907, she participated in forty-three séances with members of the IGP. In 1908, a report written by Jules Courtier was published in the bulletin of the institute. It detailed the conditions in which the séances had occurred, the results, and the conclusions of the observers. In agreeing to observe Eusapia, members of the section were hoping to verify the authenticity of the phenomena and to associate them with natural laws: "Scholars have ceased to look down on this research," he wrote. "They bring to it, on the contrary, an increasingly serious and passionate curiosity in every country. But with them, the problem has changed its face. The supernatural has been banished from it. If we immerse ourselves in the study of these phenomena, it is to check them with the rigor of experimental methods, it is to discover determinism, it is in the hope of one day associating them—if they are real—with phenomena already known and classifying them in a system of natural laws."[86] Describing the observable procedure of the experiment, he emphasized the importance of the section recording as many as possible of the phenomena produced under rigorous control and relating them to known laws.[87]

It was Courtier and Serge Youriévitch of the Russian embassy in Paris who first proposed the séances and prepared for them. With a letter of introduction from Charles Richet, both men set out for Naples, Italy, in May 1905 to meet the already famous Eusapia. In their presence, she performed eight séances in which Courtier and Youriévitch were able to observe the range of phenomena she could produce. Impressed by her abilities, the two men returned to Paris to establish a research plan for the section devoted to the study of psychical phenomena. Five different types of phenomena would be monitored and controlled: movements of objects with or without contact while recording the muscular contractions of the medium; movements of measuring instruments, such as dynamographs, or objects requiring intelligent action, such as musical instruments; action at a distance on physical instruments, such as compasses or electroscopes; action on living matter (plants or animals); and apparitions of dark or luminous forms, to be checked by means of photography and molding. To control the phenomena, instruments would be used to measure the medium's muscular movements and weight, data on the environmental conditions of the room would be recorded, and photography would provide visual proof. For example, during the experiments, observers noted meteorological conditions, electric and magnetic fields, acoustic vibrations, and x-rays, and gathered information on Eusapia's physiological state: her temperature, blood pressure, electric potential, and reflexes.[88]

The séances began on June 8, 1905. Given the fame of the medium and the prestige of the observers, they were highly publicized. Participating in the proceedings were Pierre and Marie Curie, Henri Bergson, and Arsène d'Arsonval, president of the IGP's psychical section and future president of the Académie des sciences (in 1917). Proceedings at the séances were standard: Eusapia entered the room and sat down at one of the small ends of a table inside a cabin, behind closed curtains. Her hands and feet were left visible to the participants at all times, held by two individuals sitting on each side of her. Most of the time, Eusapia rested her feet on those of her neighbors. At each séance, about five or six individuals

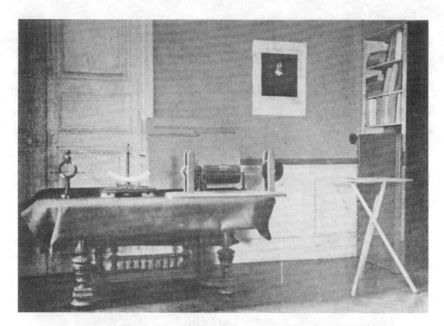

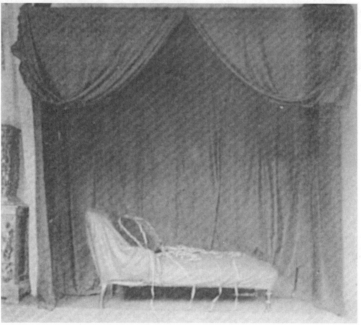

The room in which the séances with Eusapia Palladino were held, showing some of the instruments used to monitor the phenomena. While the upper photo conveys a sense of scientific observation, the lower one gives the impression of a more private setting. From Jules Courtier, "Rapport sur les séances d'Eusapia Palladino à l'Institut général psychologique en 1905, 1906, 1907 et 1908," *Bulletin de l'Institut général psychologique* 8 (1908), photos XVI and XVII.

HEURES	Lectures d'équilibre	REMARQUES
h. m. s.		
9 50	30,0	
51	30,0	
52	30,0	
53	29,6	
54	29,5	
		Saute à 24.
55	29,5	Oscillations de 26 à 31 (moy. =28,5). Mouvements du bras.
56		Osc. de 29 à 21 (m. = 25) puis, de 22 à 34 (m. = 28).
57	29,5	
		Saute à 24.
58	29,2	Osc. de 24 à 31 (m. = 27,5).
59	29,0	Osc. de 22 à 34 (m. = 28).
10 0	29,0	
1	29,4	Légères oscillations de la table.
2	29,4	La table s'agite.
3	29,6	
4	29,3	
5	29,0	
45		Légères oscillations de la table.
55		Arrêt du spot à 25,5. Soulèvement des 4 pieds.
6	29,0	
7		Oscillations nombreuses.
35		Soulèvement des 4 pieds.
50		La table s'agite.
59		Soulèvement des 4 pieds.
8	29,0	
9		Oscillations nombreuses.
35		Osc. de 24 à 30 (moy. = 27). Soulèvement des 4 pieds.
	28,4	
10	28,0	Soulèvement des 4 pieds.
35		Id.
		Osc. de 19 à 34 (m. = 26,5); léger soulèvement des 4 pieds.
11. 15		Osc. de 22 à 34 (m. = 28).
25	29,2	
		Léger soulèvement.
12		Fixe à 26.
20		Soulèvement des 4 pieds.
13		Osc. de 22 à 27 (m. = 24,5).
25	28,0	Calme.
40	27,4	Calme net et prolongé. On a détaché les fils.

Observations made with a galvanometer during one of the séances with Eusapia Palladino. The changes in electric current are associated with the movements of both the medium and the table. From Jules Courtier, "Rapport sur les séances d'Eusapia Palladino à l'Institut général psychologique en 1905, 1906, 1907 et 1908," *Bulletin de l'Institut général psychologique* 8 (1908), table I.

formed a chain around the table with their hands touching those of their neighbors.[89] At some point, Eusapia's production would begin. At first, with maximum lighting, various noises in the table would be heard. With reduced lighting, the table's feet would be raised. With still further reduced lighting, the curtain of the cabin would inflate and move. With even less lighting, objects would begin to move around the cabin. Once it was almost dark, the participants could begin to see shapeless luminescent forms moving outside but close to the cabin. At times, luminous dots and sparks also became visible.[90]

According to their reporter, Courtier, the participants never doubted that certain phenomena had been observed. The causes of such manifestations, however, had been more difficult to establish and were vigorously disputed. Through the numerous data collected with measuring instruments, the hypothesis of collective hallucination was rejected.[91] As for fraud, Courtier's report made it clear that although participants tried to impose serious monitoring, the nature of the phenomena made this difficult. Eusapia's productions often required minimal lighting, and her trances sometimes made it difficult for her to accept physical control. Moreover, controllers on either side of the medium had been asked to monitor her hand, arm, knee, and foot on that side, all with only one hand and little lighting. As the participants were both controllers and observers, they had been left in a perpetual state of divided attention and expectancy. Phenomena were never produced at the same place. They were often fleeting in character, and as a result of the long wait, the participants' attention had flagged.[92] Still other factors made serious observation difficult. Participants had been encouraged to carry on light conversation to facilitate the production. This was apparently necessary for the success of a séance, but it unfortunately created a diversion from the attempt to monitor what was happening. Also noted were other problems, including suggestibility and emotions, which could enhance the perception of the phenomena.[93]

The observers' difficulty in accepting the observed phenomena was increased by the fact that Eusapia had been caught committing fraud on a few previous occasions. At some point in 1907, for example, it was discovered that she was using a strand of her hair to move a light object toward herself. Other suspicions were formed in the cases of lifted tables. Moreover, objects moving without contact were never outside the easy reach of Eusapia; and it had been complicated to keep adequate control of her hands at all moments, Eusapia being adept at substituting her hands for those of her neighbor controller. In addition, she had refused to allow participants to take photographs without her explicit consent each time, thus rendering any attempts at monitoring in this way useless.[94]

The question of sleight of hand and fraud in mediums was never a simple one. Should Eusapia have been rejected at the first sign of it, or should it be kept in mind that, in her trances, she might lose some control over herself? In his report of the IGP séances, Courtier opted for the second course, with some reserve:

> Without going so far as to clear them [mediums]—blaming their subconscious or unconscious for their cheating would be too much, because sometimes we notice obvious premeditation to their deceptions—it should be considered that they often have a propensity to hysteria and abnormal temperaments. At séances, they lose control of themselves in their second state; in moments of hyperexcitability and tiredness, wishing to bring about the phenomena the audience is waiting for, they probably allow themselves to commit fraud, rather than produce nothing.[95]

Although participants at the IGP séances decided not to dismiss Eusapia's productions at the first indication of fraud, they did not attempt to formulate a specific explanation of the phenomena they had witnessed. Too many doubts lingered as to their origins. Here, the first step in the study of the phenomena had been accomplished, but "even had we been certain of all our observations, we would have attempted theorizing only if we could in one way or another, directly or indirectly, have tied the new facts to previously known facts, including them in the system of natural laws, because it is of this that, in the end, explaining and understanding consists."[96] Instead of an endorsement or a set of proposed explanations, the participants chose to call for further experiments with other subjects who would accept the need for control demanded by rigorous observations: "subjects who will not make us waste our time and render our efforts sterile by regrettable simulations. If these subjects exist, let them come to us, [and be] assured at the same time of finding the necessary rigor of the controls and the kindness and respect they are due."[97]

On November 30, 1908, a lengthy and lively discussion followed the reading of Courtier's report at the IGP. The idea of the institute dispatching two researchers to Naples a few months every year to experiment with Eusapia away from the crowds was considered. The medium was getting old, and her powers were declining with each passing year. It was suggested that scientists should take advantage of the remaining time during which Eusapia could perform. Funds, however, would have to be obtained in order to accomplish this. If the idea attracted some interest during the discussion, it does not appear to have been implemented.[98] Mention was also made of the possibility of young mediums being trained to work at the IGP's laboratories for a fixed annual salary. Again, the proposition was well received, but does not appear to have been implemented in the following months or years.

News of the IGP's experiment had been received with great hopes in the larger community of researchers interested in psychical phenomena. At last, a French group of distinguished scientists had agreed to take the phenomena seriously enough to observe them for some time under favorable conditions. The *Annales des sciences psychiques* showed continued interest in the proceedings and outcome of the experiments. Articles were devoted to the subject in the journal at the time, including a summary of Courtier's report in February 1909.[99] On Courtier's restrained conclusions, contributors to *Annales des sciences psychiques* could not hide their disappointment:

In the circumstances, readers of the report are naturally disappointed for the most part. Everyone is asking whether it was worthwhile keeping Eusapia at the Institut psychologique for many months, holding a respectable number of séances, inventing all sorts of recording apparatus, measuring the subject with subtle physiological instruments, photographing her from the front, from the side, and in three-quarter shots, recording her pulse and breathing, analyzing her urine daily—and spending 25,000 francs into the bargain—to arrive at such meager results.[100]

It was recognized, however, that the IGP's experiments had been useful, if only in establishing the validity of the phenomena in an objective way. At least the hypothesis of collective hallucination had been refuted once and for all, it was claimed. From the point of view of psychology, what now remained to be done was to uncover the origins of the manifestations in no uncertain terms. This would have to be accomplished through rigorous scientific control in a favorable medical or laboratory setting. Thus, although more had been hoped for, if anything, the experiments performed with Eusapia had provided scientists with a proper framework on which to build, or so many psychologists and psychical researchers thought.[101]

Whereas Courtier had remained fairly restrained in his report of 1908, writing in the name of a group of scientists, he did commit himself further in a postscript written in 1928. Although he confirmed his conviction that the experiments with Eusapia were to be understood as the foundations for further observations of the phenomena and could not be taken, by themselves, as sufficient proof of the authenticity of such manifestations, he did make explicit his own personal conviction that he had not been deceived, and that some of the phenomena he had witnessed had, in fact, been real. If his report in 1908 had been meant as a collection of facts that required the reader to form his or her own conclusions, he was able, twenty years later, to look back on the séances with Eusapia and affirm his belief that what he had witnessed at the time had been authentic mediumistic phenomena, especially those involving moving tables without action. For Courtier, the IGP's 1908 report gave probable cause for even skeptics to accept the existence of such phenomena.[102]

Although conceived initially as an institute dedicated to the study of the supernatural, the IGP quickly evolved into a more general organization. The Section des recherches psychiques et physiologiques remained an important group in the institute throughout the 1910s and the 1920s. On occasions, the control of mediumistic phenomena, divining rod experiments and observations, telepathy, and clairvoyance continued to be discussed.[103] No considerable project witnessed by distinguished physicians and scientists like the one with Eusapia was ever organized again, however. The last published bulletin of the IGP appeared in 1929.

Little remained by then of the initial agenda of the institute as a scientific institution dedicated to the study of mediumistic phenomena and other occurrences of the kind. Over the years, the initial curiosity about séances had died down. It was not that members had succeeded in explaining mediumistic phenomena to their own and one another's satisfaction, but rather that they had lost interest in or given up on a set of manifestations that could not be easily incorporated into psychology.

ଇଚ

The gradual dismissal of the supernatural by the IGP reflects a general attitude in the medical and the human sciences in the first decades of the twentieth century. By the nature of their work on the human mind, both in its normal and its pathological manifestations, physicians, psychiatrists, and psychologists had come to consider mediumistic and similar phenomena. They had attempted to explain them using the concepts and theories of their fields. More than any other groups, they had been able to bring a certain degree of respectability to claims of supernatural occurrences, but they had done so at a high cost in the eyes of all those who believed in them. In hospitals wards and psychological laboratories, mediums, stigmatics, and visionaries had become patients and subjects of research. Those who believed otherwise—séance-goers and other followers—were apparently deluding themselves. If physicians, psychiatrists and psychologists claimed to have sufficiently explained the causes of supernatural occurrences of these kinds, however, their success was never complete. Outside a few works in the context of personality studies and some vague mention of pathological conditions, no clear and definite theory was ever formally accepted. Members of each group who had shown an interest in the phenomena had often felt discouraged by their lack of control over the proceedings and deterred from pursuing this line of research further. For those who persisted, the outcome was more often than not perplexing, and the conclusions were rarely definitive. In the end, physicians, psychiatrists, and psychologists were poorly equipped to deal with the fleeting and elusive manifestations of the supernatural. Their conception of mediums and others who experienced the supernatural as subjects of research or patients limited their work. In contrast, psychical researchers accepted that the investigation of supernatural phenomena demanded a more flexible approach—assuming they were real—and would attempt to develop methods of observation and control in collaboration with mediums and other subjects.

CHAPTER FOUR

Witnessing Psychical Phenomena

Psychical research officially began in France with the founding of the *Annales des sciences psychiques* in 1891. In the opening pages of the first issue, the physiologist Charles Richet declared: "We are of the firm conviction that, alongside the known and described forces, there are forces that we do not know; that ordinary, simple mechanical explanations do not suffice to explain all that is happening around us."[1] Over the next three decades, he contributed to the *Annales* while continuing to hold a post as professor of physiology at the Sorbonne. Although the *Annales* was not the only journal devoted to psychical research in France at the time, it was the central one, meant to represent the field with accuracy and integrity. Until its demise in 1919, when the study of psychical phenomena took a new direction as metapsychics, psychical research in France developed almost entirely around the journal.

Psychical researchers were men and women who situated themselves between spiritists and occultists on one side and physicians and psychologists on the other. They believed that the phenomena witnessed at séances were authentic, but that they were caused by unknown mental abilities rather than spirits. They also thought that séances were natural and the product of gifted individuals, not mental pathologies. They considered themselves to be broad-minded, objective, and free from any assumptions, spiritual, scientific or otherwise. Their mission was ambitious: to provide psychical manifestations with a scientific status. Psychical research would be a science in its presentation and its rigor, but it would be a different kind of science. It would be accessible and inclusive: an open enterprise. Anyone could subscribe to the *Annales*, which was aimed at a broad audience rather than a restricted group with specific disciplinary credentials. Psychical phenomena could be observed or experienced by everyone, and readers were encouraged to participate by reporting their own psychical experiences to the *Annales*. In this way, psychical research differentiated itself from other scientific disciplines, which had become increasingly exclusive during this period. In a number of manners, psychical researchers were reacting to the growing professionalization of the scientific disciplines, and the relegation of the public to popularized versions of the sciences, by offering the promise of a new field that would develop alongside the public's increased fascination with scientific wonders.

THE MOVEMENT AND ITS ORGANIZATION

Psychical research originated in Britain and formally began with the 1882 creation of the Society for Psychical Research (SPR), the first society founded for the sole purpose of observing psychical phenomena. Attempts to provide a scientific basis for the phenomena witnessed at séances can, however, be traced to earlier years. In 1869, the London Dialectical Society had created a commission of thirty-three members to study and experiment with psychical phenomena. Their report had deemed the phenomena worthy of investigation. In the early 1870s, the chemist William Crookes began to experiment with the mediums D. D. Homes and Florence Cook. Then, in 1882, William Barrett, professor of physics at the Royal College of Science in Dublin, and Edmund Dawson Rogers, a press agency manager and amateur naturalist, set out to create a serious research society that would overcome the image of charlatanism often associated with the phenomena. Henry Sidgwick, professor of moral philosophy at Cambridge, was asked to be the SPR's first president.

Sidgwick's distinguished reputation and his scholarly interest in developing a rational concept of the soul made him an appealing choice for the new society. A governing council, and a president elected by it, constituted the leadership of the SPR. The research itself would be accomplished through investigating committees: one on thought transference, rapidly redefined as telepathy, one on mesmerism, one on Karl Ludwig von Reichenbach's notion of an Odic force, one on haunted houses, one on spiritualism, and one to collect historical material and contemporary testimony in all related areas.[2] In 1884, the SPR began publishing the *Journal of the Society for Psychical Research*, which still exists today. That same year, the American Society for Psychical Research was created; it would become an affiliate of the SPR after 1889 due to financial difficulties.

In 1886, the SPR produced an extensive study of hallucinations, apparitions, and telepathy. Edmund Gurney, Frederic Myers, and Frank Podmore's *Phantasms of the Living* was a collection of 702 cases, the majority of which were reports of sightings of an individual either twelve hours prior to or following death. *Phantasms of the Living* became a landmark in the field of psychical research as the clearest defense of the scientific validity of the phenomena to date. By choosing to discuss visions of the near-dead, the SPR was beginning to distance itself from the spiritualists who focused on communications with the dead as opposed to sightings of the near-dead. Psychical research was becoming a separate field of inquiry, more preoccupied with telepathic phenomena than communications with spirits.[3]

In France, the more scientific branch of spiritism took longer to develop; and it was not until the creation of the *Annales* that French psychical research really took shape. Lack of structure did not mean lack of interest, however; and the work of the SPR was often discussed in French societies during the 1880s. France was also prominently represented in the membership of the SPR throughout the decade.[4] But if some were familiar with psychical research at the time, and even approved of it, no one in France seems to have referred to themselves as psychical researcher before 1891. Rather, any sympathetic consideration of the phenomena outside of spiritist and occultist circles seems to have remained entrenched in a tradition of physiological psychology that would continue to influence the French version of the field later on. In fact, it was Charles Richet, one of France's most distinguished physiologists, who pushed for the creation of a French journal of psychical research. Richet had been a professor of physiology at the Faculté de medicine in Paris for a few years already when he created the *Annales*. He would go on to have a tremendously successful career as a member of the Académie de médecine starting in 1898, a Nobel Prize winner in 1913 for his work on anaphylaxis, and a member of the Académie des sciences starting in 1914. The *Annales*, as Richet introduced it, however, would be dedicated to his less orthodox interest. It would focus on the phenomena of telepathy, lucidity, and premonition. Observations of such occurrences, he wrote, would reveal an unknown ability of the human soul. The journal would also consider physical phenomena of all kinds including objects moving without contact or ghosts and apparitions, although he himself remained doubtful of their existence.[5]

The *Annales* appeared every two months from 1891 to the Great War, when it continued to exist with reduced productivity until its last issue in 1919. In its first year of existence, it was mostly used to relay observations sent by readers and discuss the methods and approaches required to observe and understand the phenomena. Psychical research was too young to produce hypotheses, Richet asserted in the journal's first issue. For now, the *Annales* would focus on observations rather than theories, which would come later, once the phenomena had been sufficiently observed and confirmed. The science was young and in its empirical period, filled with scattered observations: "Let us resign ourselves thus to be mostly observers and not experimenters"[6] With this agenda, Richet spearheaded the first commission dedicated to the study of telepathy. He invited journal subscribers to send in reports of their own experiences with psychical phenomena.[7] A year later, Félix Alcan, publisher of the *Annales* and numerous other works at the time, explained that so far, the journal had limited its focus to telepathy, but that it was his hope that, as reliable reports began to reach it and the reality of the phenomena could be established with greater certainty, the journal would widen its concerns to the study of other interesting manifestations.[8]

In 1901, as editor of the *Annales*, the physician Xavier Dariex reflected that in the past decade the journal had succeeded in its original mission, which was to examine psychical phenomena impartially and report on them credibly to the scientific world, while avoiding theoretical discussion.[9] The *Annales* was now ready to enter a new phase of its existence: "we will abandon our strictness—our exclusiveness as some impatient minds would say—and widen our program."[10] This redefinition of the domain of the *Annales* was not the only change in French psychical research at the beginning of the twentieth century. More psychical journals and societies were beginning to appear around the country, each one with its own approach. Lyon, Marseille, Nancy, and Nice all had their own societies of psychical research, each focused on their own region and taking its own stance on the phenomena. Whereas the Société d'études psychiques de Nancy affirmed doctrinal neutrality and limited its domain of inquiry to collecting reports and assessing their authenticity, the Société d'études psychiques de Nice shared in the occultist interest of many inhabitants of the region.[11] If more journals and societies flourished during the first decades of the century, however, *Annales* continued to set the tone and to provide a focus for the French movement.

In 1904, the *Annales* grew in importance when it incorporated the *Revue des études psychiques*, a journal initially created by the criminologist Cesare Lombroso in tandem with a journal of similar name in Italy, the *Rivista di studi psychici*. This enabled the *Annales* to increase its publication rate—it began to appear every month, as opposed to every two months as had previously been the case. In addition, the content of the journal increased significantly. Dariex continued to be editor of the journal and César de Vesme, formerly editor of the *Revue des études psychiques*, became the *Annales's* editor-in-chief.[12] A new editorial committee was formed consisting of Camille Flammarion, Émile Mangin, Joseph Maxwell, and Albert de Rochas from France, and William Crookes, Cesare Lombroso, Enrico Morselli, Julien Ochorowicz, François Porro, and Albert Von Schrenck-Notzing from abroad. Richet was not part of the official committee, but he was said to provide assistance to its members. The fusion of the two journals brought other changes to the *Annales*. Formerly presented as a journal of observation and, later on, of observation and experimentation, it now featured information and news of the psychical movement both in France and internationally. More than ever, it assumed a central role in the movement.[13]

The *Annales* continued to evolve. In 1908, it became the official publication of the Société universelle d'études psychiques (SUEP) founded in 1901 by the physician Paul Joire in Lille and later relocated to Paris.[14] Joire had wished to create a society in which psychical phenomena would be studied scientifically. The SUEP did not represent any doctrine, and membership in the society did not imply belief in the authenticity of the phenomena. For Joire, his was a scientific society, and meetings would be conducted as such: "The Society absolutely forbids any discussion outside those of purely scientific questions," he wrote.[15] The SUEP had three objectives: the scientific study of mediumistic productions, telepathy, hauntings, lucidity, predictions, apparitions, materializations, and other related phenomena through experimentation, debates, lectures, publications, and contests; the propagation of this knowledge to the public; and the promotion of the use of scientific methods in the field.[16] Starting in 1908, the reports of the SUEP were published in the *Annales*. Members of the SUEP now received the journal, and any subscriber to the journal was allowed to participate in the activities of the Paris section of the society.[17] The association with the SUEP marked a new direction for the *Annales*. Accounts of meetings, lectures, and experiments of the SUEP now occupied a significant portion of the journal, confirming its position as the most important journal of psychical research in France.

That same year, the *Annales* became an illustrated, biweekly periodical. "[P]hotography has taken, of late, such a documentary importance, above all in mediumistic research," wrote the editors. "[T]he drawings of apparatuses, the plans, and the diagrams of autographical tracings play such a prominent role in most experiments that, after having struggled with the technical difficulties resulting from the limited proportions of our format, the quality of our paper, etc. . . . , we have agreed on the radical reform that we have just announced."[18] The rapid changes seemed to have caused anxiety in some of the journal's readers. At the beginning of 1908, Vesme wrote: "We will not insist on the petty prejudices that those who find themselves bothered in their habits harbor against the improvements to the *Annales*, and we have complete confidence that the support of those devoted to the study of psychical sciences will allow us to accomplish the tiresome but so useful task that we have taken upon ourselves."[19]

With its new format and its association with the SUEP, the *Annales* had become even more accessible and spectacular. Its pages were now filled with evocative pictures and illustrations and an increasingly vast domain of interest. Gone were the days when the journal limited most of its activities to the sober discussions of telepathic occurrences; it had become a truly popular enterprise. The more accessible it became, however, the less likely it was to be taken seriously by the scientific world. Ultimately, the creators of the *Annales* failed in the mission they had set themselves. In 1911, the journal's founding editor resigned from his position. Frail and having lived outside Paris for a few years already, Dariex sensed he had become "the shadow of an editor that is not yet of the other world but that no longer appears to be of this one."[20] He felt he was no longer able to provide the journal with the kind of attention it deserved. After twenty years of involvement, Dariex left the *Annales* behind. During the Great War, the journal understandably slowed down its activities. Only a few issues appeared between 1914 and 1918. In 1919, then *Annales* published its last issue. By then, the Institut métapsychique international (IMI) had just been created and took over the affairs of the *Annales* and the SUEP.[21] With the end of the war, psychical research was being replaced by metapsychics in France.

A PARTICIPATING PUBLIC

Psychical research allowed the public to become more than passive witnesses to scientific progress. Members of the general public could join scientists in the development of the field. By collecting testimony, writing accounts of séances, proposing a classification for psychical experiences, or suggesting ways in which to account for the phenomena, anyone could become a psychical researcher. Even women were encouraged to give personal accounts of psychical experiences, publish their research, or present scientific papers at conferences. One of the most renowned psychical researchers of the 1910s was Juliette Bisson, famous for her work on ectoplasmic productions. Although academic credentials were always welcomed and even overblown, books on the topic were read and discussed whether or not their authors had diplomas or prestigious affiliations. Many psychical researchers began their accounts with humble remarks on their lack of experiences in the sciences. For example, in a work on thought transference and clairvoyance, A. Bonneville wrote: "I implore my readers to be lenient. . . . I left elementary school very young. I was faced with the difficulties of life early on, and

I came to my present situation only by dint of hard work and energy."[22] Others expressed frustration at the restrictive nature of the scientific enterprise, feeling that the academies and universities had taken control of scientific investigation. For example, the rector of the Académie de Dijon, Émile Boirac, deplored that science had acquired the authority of dogma and that those who did not possess its knowledge simply had to accept it without much choice.[23]

Religious imagery and analogies were common and evident in the vocabulary used by members of the movement, even though many were not devout Roman Catholics. Psychical researchers described a world of science in which very little room was left for novelty or tolerance. A Bordeaux lawyer and author of a few works on psychical phenomena, Joseph Maxwell, went so far as to compare scientists to "the ecclesiastical authorities of the Middle Ages." Scientists today "treat independent scientific thought as the inquisitors treated free thought long ago. They possess the same intolerance, the same hatred of schism and heresy, as their prototypes of former times."[24] For all their outrage, however, psychical researchers held an ambivalent attitude toward science, and they seemed to regard themselves as outsiders. On the one hand, they disapproved of its restrictive character and promoted a more accessible kind of enterprise. On the other, they were ardent defenders of the need to adopt a scientific approach in the study of psychical phenomena and wished their field to be accepted as truly scientific, with the authority and prestige that came with that standing. In that sense, like spiritists and occultists, psychical researchers picked and chose when to attack science and when to embrace it depending on their interests at a specific moment in time.

Psychical researchers did not limit themselves to the study of mediums or other subjects they could observe first-hand. Rather, it was a fundamental principle of the burgeoning field that psychical phenomena were democratic; they could be experienced by anyone. "Psychical phenomena occur everywhere: they are not the privilege of any social class: they can just as well be observed in the most humble cottage as in the most sumptuous palace," wrote Dariex in the *Annales*.[25] Psychical phenomena were neither rare nor to be seen as the morbid manifestations of a diseased mind; rather, they were the signs of unexplored and unexploited potential; abilities present in all human beings. Across France, societies of psychical research welcomed contributions of any sort from the inhabitants of their region. Psychical research was a science developed through public participation and accessible to anyone. Launching the *Annales* in 1891, Dariex called upon his readers to pay close attention and inform the journal of any telepathic occurrence they might experience or witness. He explained to them how to produce a report so as to ensure its reliability. Readers were to write to the journal of any event, even those that appeared without interest, but to take every precaution in this. Memory was unreliable, and reports should be corroborated by the testimony of others as far as possible.[26]

If no definite proof of the existence of psychical abilities in the population had been found so far, it was probably because very few individuals knew how to channel their powers. Around the beginning of the twentieth century, a plethora of brochures and books by various unknown authors claimed to teach their readers to develop such powers through diverse methods. Henri Durville's *La télépsychie*, for example, was written as a course on telepathy in which readers began by learning how to perform muscular readings and progressed in their abilities with each lesson. Anyone could learn how to use their sixth sense, the author promised. With the techniques presented in *La télépsychie* and with serious practice, readers could learn enough to dazzle family and friends. First, they had to become aware of how thoughts translated into unconscious acts. By holding someone's hand, they would learn to detect subtle movements and thus develop the mental concentration and sensibility necessary for thought reading. Once this muscular reading was perfected, thought reading could be practiced. First with cards, then numbers, and finally sentences, readers were told how to uncover this buried faculty of the human mind.[27]

Authors promised that such powers would bring happiness, success, and popularity to their readers. In a course on hypnotism, one author discussed how it could be used at a séance for amusement.[28] Another emphasized how thought reading and suggestion would bring power to their practitioners, allowing them to attract and seduce others or prompt absolute fidelity in a lover. The brochure was accompanied by an object called a *radio-suggestif*, which was said to enhance the customer's personal influence or psychic fluid with its magical and suggestive action.[29] Another author developed a course on the ability to influence others, even from afar. By repetitive efforts to bring a particular thought into his subject's mind, the practitioner could slowly permeate the unconscious depth of an individual, bringing about new ideas. This could be used to influence, cure, protect others, perhaps even inspire love in a desired person.[30]

Psychic influence was not always directed at controlling other people. More than a spectacular parlor trick, telepathy could also bring about good, Durville reassured his readers. There was nothing to fear; all this was perfectly natural. In fact, it had probably been the initial means of communication in primitive populations.[31] Some brochures claimed that mental forces or the power of autosuggestion could lead to a healthy and successful life.[32] Others asserted that certain exercises could be practiced to retain, even increase, one's fluidic force and use it to promote health and psychical powers.[33] Still others affirmed that an occult force existing throughout the universe could be channeled by individuals with a sufficiently developed magnetic fluid. It would allow them to perform suggestive therapies and grant them the power to cure headaches, toothaches, and migraines. "Disease itself does not exist, the pain that we suffer is only the result of an organic disagreement, that is to say a lack of harmony, a lack of vitality, it is the disequilibrium of the vital movements of the organism."[34] Other works discussed the curative powers of a force comparable to human radioactivity and claimed that such therapeutic approach lay in the mental perception of the colors produced by the vibrations of cosmic currents. By harmonizing their aura to these currents, individuals could appropriate the thoughts and desires that manifested themselves in such a current. Different colors could be used for different purposes. Green, for example, could be used to gain money or against cancer.[35]

For psychical researchers, the readers of such work could provide material for investigations. The gathering of testimony was essential to the field. The earliest collection of reports from the population dated back to 1886 when Frederic Myers, Edmund Guyers, and Frank Podmore published a two-volume set of case studies titled *Phantasms of the Living*. In 1899, Flammarion undertook a similar project in France when he asked the readers of the *Annales politiques et littéraires* to report on any personal experience for which they could not account through the known laws of nature. The survey yielded four thousand answers describing a variety of phenomena that the astronomer classified into sixteen categories: manifestations of the dead, manifestations of the dying, manifestations of the living, visions of faraway events, dreams about the future, dreams anticipating death, meetings foreseen, clairvoyance, double of a living being, telepathy, impression given by animals, calls heard from great distances, telekinesis, doors opening alone, haunted houses, and spiritist experiences.[36] In his 1900 book *L'inconnu et les problèmes psychiques*, Flammarion reproduced a number of the testimony he had received and discussed most of these manifestations of the soul. By then, he had long abandoned the spiritist hypothesis and chosen to focus most of his attention on thought transference (telepathy) as experienced by the general population, especially as it related to the moment of death. He explained it using a concept of cerebral vibrations. Each sensation or idea, whether a waking thought or a dreaming one, corresponded to a cerebral vibration, a movement of the cerebral molecules. For telepathy to occur, two minds had to be in harmony and vibrate in synchronism. Flammarion argued that telepathy, if shown to exist, would prove the existence of a soul independent of the body. Although not a proof of immortality just yet, it would be a step in the right direction.[37] During the Great War,

Richet launched his own public investigation when he asked soldiers and their families to send in reports of any premonition of death, whether confirmed or not. As a result, the 1919 issue of the *Annales* announced that premonitions of death had occurred more frequently during the war than before.[38]

At the basis of these large-scale collections of testimony was the very simple conviction that the greater the evidence, the greater the likelihood of a phenomenon's authenticity. Methodologically, this reliance on public testimony was problematic. Psychical researchers worked with the observations of untrained individuals. How could an investigation aiming for scientific status be based on unreliable observers? In the first issue of the *Annales*, Richet discussed this particular challenge of the field. The problem was one of method, he felt. Testimonies were too often hazy and questionable. One had to wonder about the sincerity of the observer: Could this person be trusted? Was the witness sick? What was their mental state? Did the witness experience frequent hallucinations? Could others confirm the experience? In the case of telepathy, had the times of transmission and reception been recorded and did they in fact correspond? Finally, in a case of manifestations of the dying or the near-dead, could official documents be provided to confirm time of death?[39] Generally, the decision to accept or reject a report was based on reputation and education. In *L'inconnu et les problèmes psychiques*, Flammarion had vouched for the credibility of witnesses of his acquaintance. He felt they were individuals of honor.[40] In fact, many psychical researchers believed the bonds between honor and credibility to be strong enough to justify the use or the rejection of a testimony. In the context of late nineteenth-century France, this hardly appears surprising. The historians William Reddy and Robert Nye have both written about honor as a central component of French society during this period, arguing that it became to the new bourgeoisie what birthright had been to the aristocracy of the previous eras.[41]

Problems with these particular types of observations ran even deeper. The spectacular nature of psychical phenomena could make even the most honorable of observers lose their judgement. Richet warned that "exquisite tact" was required to avoid being seduced by appearances. "A good experiment in objective metapsychics is extremely difficult," he added. It was necessary to be suspicious of everything and of everyone, and above all of ourselves. "Our intense desire to see the experiment succeed must not lead us to deceive ourselves."[42] The metapsychist René Sudre would later stress the observer's responsibility to remain unaffected in the face of the extraordinary: "He must not be impressionable, or at least he must be able to control himself and to keep a clear mind. He has to be benevolent and at the same time distrustful and not very gullible."[43] Exaggerations and modifications of the events were frequent when remembering a striking experience. For Richet, a great number of the reports were unreliable for such reasons. It was not that witnesses were lying, but that they could not relay an experience adequately.[44]

Furthermore, even in the cases in which an occurrence had been reported faithfully, there was always the danger that it had been caused by a hallucination or an illusion. Richet never neglected this possibility.[45] Others agreed. For Boirac, the two principal dangers of a science relying on testimony were illusion and simulation, whether conscious or unconscious. For psychical research to become scientific, it would have to be purged of both. Observers would have to be critical of their own experiences at all times.[46] On these same dangers, Maxwell remarked: "I have to defend myself only against two enemies, the fraud of others or my own illusions."[47] Flammarion was also aware that hallucinations or even coincidences could explain many reports. Considering these possibilities, he asked his audience to send him reports of any premonition, even those that had remained unfulfilled. In his survey, only seven or eight cases out of every hundred experiences relayed had not been followed by a death. The argument followed that if premonition had always been caused by coincidence or hallucination, the incidence of unsuccessful cases would have been much higher. For Flammarion, this constituted sufficient proof against these threats to reliability. He later wrote: "Their reality [the phenomena] is mathematically proven by the calculus of probability."[48] He was not the only one to argue on the basis of probability. Richet discussed the advantages and the limits associated with such a tool. Psychical experiments could not be reproduced at will. A medium who had produced wonderful phenomena in a series of séances could cease to perform at any given time. Thus, in many cases, the phenomena did not lend themselves well to the use of probabilities, but it remained that the greater the number of positive results, the greater the likelihood of authenticity.[49]

If Richet readily admitted that only a few of the innumerable documented cases could be taken seriously, and, if he agreed that the quantity of unreliable testimony was overwhelming, he nonetheless believed that there had been enough trustworthy testimony to infer the reality of the phenomena: "The authority and the repetition of testimony and proofs does not allow us to doubt any longer. Cryptesthesia, telekinesis, ectoplasmy, and premonitions appear to me to be established on bases of granite now"[50] There were some who disagreed. Even amongst psychical researchers, there were a few who continued to question the validity of a science relying on stories reported by the public, understanding nonetheless that to reject such stories would cripple the field.[51] In the end, however, by its nature, the field of psychical research had to rely on the testimony of untrained witnesses, often grieving relatives and friends in difficult observational settings. Problems were and would always remain unavoidable.

All of this did not go unnoticed outside the psychical research movement. In fact, psychical researchers were often attacked for their use of personal testimony. Paul Heuzé, a journalist at the weekly newspaper *L'opinion* and one of the most outspoken adversaries of psychical researchers, denounced the methods of the field in numerous articles and books. Heuzé was not entirely opposed to the possibility that telepathy or other psychical manifestations of the intellectual kind existed, but he was very critical of the reliance on the ill-educated general population for witnessing. He believed that, when it came to testimony, psychical researchers often favored quantity over quality. Flammarion was particularly guilty of this, he asserted. The cases he discussed were often ancient or vague, and some of the stories crumbled or evaporated when Heuzé attempted to authenticate them.[52] "[I]t is pieces of gossip that we are offered, a lot more than documents," he deplored.[53] Even the great Richet was not beyond reproach, according to the journalist. His naïveté regarding what he perceived as the "good faith" of certain individuals "above all suspicion" was often apparent. For Heuzé, such declarations were uncritical to say the least, if not deeply suspicious.[54] For psychical researchers, however, reliance on witnesses was unavoidable given the nature of their field and its phenomena.

IN THE PSYCHICAL RESEARCH PRESS

Journals like the *Annales des sciences psychiques* did not read like dry science periodicals. On the contrary, psychical research journals were made to be interesting and attract a wide audience. News and events said to involve special abilities were reported and followed from one issue to another. Among others, many court cases in which claims of the supernatural had been made by the defense featured prominently in the press of the field. Over the centuries, the supernatural had at times found its way into the courtrooms. This tendency continued even at the beginning of the twentieth century. Occasionally, the courts were asked to rule on the validity of spiritist manifestations. Cases of this kind interested both psychical researchers and the public and were often discussed in the movement. In particular, charges of fraud, attempts to control others by supernatural means, and claims of hauntings filled the pages of psychical research journals.

In some instances, mediums and others claiming extraordinary abilities faced the court on accusations of fraud. The most famous of spiritist trials was the 1875

Buguet trial on spirit photography, but there were others. In 1906, for example, a young woman known as the clairvoyant of Saint-Quentin was charged with fraud for practicing medicine without a license or credentials. Her brother and father were also charged with fraud for acting as her magnetizers. Estelle Bar had claimed that, in a state of suggestion, she possessed the ability to diagnose and cure diseases, even from a distance. The defense asked the expert psychologist Paul Magnin to verify Bar's claim. Although Magnin confirmed that the accused could be put in a state of hypnosis, he did not think that such a state implied the diagnostic abilities Bar had claimed.[55] Faced with the damaging report of their expert witness, the defense asked for a second opinion. The physician Hippolyte Baraduc was asked to give his own views on Bar's clairvoyant abilities. He reported: "This young person appreciates the degree of vitality of the sick organs without knowing their anatomo-pathogenic nature. . . . She is in contact with the biological dynamism of the organs; the radioactivity of the sick organ and her personal psychometrical degree are the means and the instruments of her work."[56] For Baraduc, Bar was not a superior clairvoyant. She was what he called an unconscious psychometer, sensing the "pathogenic vibrations" emanating from the "radioactivity" of "sick organs." With this report, it was decided that the clairvoyant, her brother, and her father had all practiced in good faith, and the defense was able to have the charges of fraud dropped. The charges of illegal practice of medicine, however, remained, and the accused were declared guilty on this count and fined one hundred and forty francs.[57]

The possibilities and limitations associated with hypnotism had been fought over in academia and the courtrooms of France throughout the 1880s and 1890s. At the time, two groups, one consisting of physicians at the Salpêtrière hospital in Paris and the other of physicians, physiologists, forensic experts, and lawyers in Nancy, had argued over the nature of the phenomenon. In Nancy, Hippolyte Bernheim, Ambroise Liébeault, Henri-Étienne Beaunis, and Jules Liégeois had believed that anyone could be made to act unconsciously through suggestion. In Paris, a group of physicians around neurologist Jean-Martin Charcot had argued that suggestion only occurred in individuals with tendencies to hysteria. In 1890, the two schools faced each other in the courtroom at the trial of Gabrielle Bompard, a woman accused of complicity in a murder and a robbery. In her defense, Bompard's lawyer argued that she had not been in control of her actions at the time of the crime, having acted under a posthypnotic suggestion induced by her partner. In this, he was supported by the expert testimony of Jules Liégeois, professor of administrative law at Nancy. Liégeois failed to convince the court, and Bompard received a sentence of twenty years for her crime.[58]

Instances such as the Bompard case in which control over others was used as a defense in court were of particular interest in the psychical research press. In 1901, for example, the *Moniteur des études psychiques* reported on two court cases, one in Leipzig at the Supreme Court of the German Empire and one in Liège, Belgium, both addressing the question of whether to allow the defense to use spiritism in a murder case. In Leipzig, the court declared that the attempt to kill using incantations or spiritist means could not be punishable by the code. Since intent to harm was not a crime, neither were attempts to harm that had occurred outside the realm of physical and psychical causality.[59] In contrast, in Liège, a case for which the defense used the notion of irresponsibility due to magnetic influence was heard. After a fight, Annette Andrien, under the influence of alcohol, had shot and killed her lover. At the trial, the defense discussed the victim's magnetic hold on the defendant. Appealing to an article of the criminal code stating that no infraction was committed if the perpetrator of the crime had been in a state of insanity or under the influence of forces that they could not resist at the time of the act, the defense pleaded that the death had, in fact, been a suicide. Having lost his desire for life, the victim had used his magnetic abilities to suggest the killing to his lover. To the dismay of the prosecutor, who argued that even under hypnosis a person did not lose the ability to choose, an impressed jury acquitted the young lady.[60] In 1909, another case of this kind was heard by the tribunal of the Seine. Madame Lobs had left all of her possessions to her daughter aside from a specific bequest to her young medium. The daughter asked the court to nullify this bequest, attributing it to a "spiritist maneuver" on the part of the medium to circumvent her mother's will. The tribunal disagreed and decided that spiritist practices did not suffice to establish insanity; the will was thus valid.[61]

Judicial cases of hauntings also occasionally came up in the psychical research press. The *Annales* and other journals debated a number of legal issues. For example: Could a lease be broken when a claim of haunting had been made? What kind of evidence would allow for a contract to be broken? These questions had at times been presented to some of the courts of Europe. In France, in 1576, the *parlement* of Paris had ruled against the breaking of a lease following claims of hauntings. In this particular case, the authenticity of the haunting had not been disputed, but it had been understood as a religious question and not for the courts to decide. Whereas in Paris, the *parlement* usually had ruled to honor contracts in cases of hauntings, in the provinces, courts had tended to rule that apparitions were a sufficient cause to break them. After the French Revolution, however, magistrates from across the country became increasingly reluctant to accept cases of hauntings. For the spiritist César de Vesme, this would change if psychical researchers were called in to provide their expertise and help to differentiate between frauds and genuine cases of hauntings and poltergeists. The marvelous was now ready to enter French case law in a rigorous manner, he claimed.[62]

A few psychical researchers and enthusiasts tried to formulate an explanation for cases of hauntings and poltergeists. The Belgian playwright and novelist Maurice Maeterlinck explained hauntings, or "unknown guests," as caused by the same spirits of the deceased that had turned tables and provoked automatic writings. If the messages were confused, he claimed, it was because neither séances nor hauntings were settings in which clear connections could be established between the living and the dead.[63] Flammarion also collected stories of hauntings in a book published on the subject in 1923. Haunted houses were widespread, he claimed. A number of people had been so terrified by the apparitions in their homes that they had abandoned their property forever. Hauntings were caused by different entities. Those associated with the dead were the product of a force inherent in the soul and acting physically in ways comparable to electrical disturbances in electricity or vibrations in the ether. Hauntings could also be caused by unknown invisible forces, or by the souls of animals or idiots.[64] Finally, some cases could be explained by matter, which had the property of registering or preserving vibrations and emanations, be they physical, psychical, or vital. He called this capacity of the walls and furniture to impregnate themselves with vibrations and give out auras "telesthesia."[65] Hauntings were thus complex phenomena, difficult to explain, but nonetheless real, he asserted.

Émile Tizané, a retired French policeman who dedicated his free time to the investigation of claims of the marvelous such as Marian apparitions and hauntings, used Maeterlinck's concept of the unknown guest to explain the many phenomena he had encountered in his work. Because his position as a lieutenant allowed him easy access to most police records on the subject, Tizané was able to assemble as much evidence as possible on a variety of cases. Like most people, the policeman had heard about and read some of the claims of hauntings that had been made in the press over the years. In the 1930s, he began to investigate certain newspaper accounts more closely. In particular, he obtained the police reports that had been written in the cases in which hauntings had caused damages or disturbances and had required some sort of intervention. Digging around, he found that most incidents of hauntings reported in the press had been solved by the police. In most cases, investigators had rapidly uncovered a fraud and its perpetrators, usually young men or women. More often than not, however, this conclusion to the affair had not made it into the newspapers. The press, which had been interested by the initial sensational news, had subsequently lost interest, giving readers the impression that there were a lot more occurrences of unexplained hauntings than was in fact the case. Tizané did find, however, that a few cases of haunting had remained unsolved. Focusing on such cases, he became puzzled by events he could not explain without accepting the existence of some sort of unseen force, the unknown guest, which he described as an invisible being capable of acting on its own, in close relation to a dynamic element belonging to a living organism, or by directing the actions of a human being. Tizané believed that scientists would soon learn to control the unknown host and direct its actions.[66]

Whatever the conclusion, investigating hauntings was not without its problems. When considering such phenomena, most stories were dated and often erroneous. Richet showed very little faith in cases recorded in the past and chose to concentrate on contemporary occurrences. In the case of haunted houses, he recognized that the limited possibilities of experimentation meant that the focus would have to be on observation and detailed consideration of the testimony.[67] An examination of hauntings led him to believe that ghosts did appear at times and to many individuals, either successively or collectively, and that noises and movements of objects in some houses defied rational explanations. First eliminating fraudulent hauntings, he suggested that most of the remaining cases were caused by spontaneous telekinesis, the unconscious ability in some individuals to move objects at a distance. This was still a hypothesis, he noted. The phenomenon required further investigation before it could be more firmly established as the potential cause of hauntings.[68]

Alongside reports of court cases and claims of hauntings, animal wonders also gathered significant attention in the psychical research press. If humans possessed unknown psychical abilities, could the same be said of animals? Could psychical researchers uncover a way to communicate across species? Could they provide evidence of a higher intelligence in animals? Within the new evolutionary framework and its breakdown of the demarcation between humanity and the animal kingdom, such questions were now conceivable. In the first two decades of the twentieth century, discussions of animals and their intellectual and psychical abilities fascinated psychical researchers and psychologists alike. In the psychical research press, claims of animal abilities were always popular.[69]

In particular, accounts of three calculating horses from Germany captivated readers of psychical research journals. At the turn of the century, in Elberfeld (now Wuppertal), the trainer Wilhelm von Osten claimed to have succeeded in training his horse Hans to count out numbers he had heard by tapping with a hoof, perform simple arithmetic tasks, tell the date, and even read. The phenomenon caused a great stir. Psychologists and physiologists from Berlin came to observe the horse. After prolonged experiments with Hans, Oskar Pfungst, a student of psychology in Berlin, concluded that the phenomenon was easily explainable: the trainer was giving Hans the answers through small unconscious movements of the head or the eyes. In 1907, Pfungst published his findings. The phenomenon was considered understood; the matter appeared closed.

In 1912, the publication of a new book, *Denkende Tiere* by Karl Krall, ignited a renewed interest in Hans and other animal wonders. Krall was a rich and respected merchant of Elberfeld. Upon von Osten's death in 1909, he had inherited Hans and decided to continue the horse's education. He claimed that Hans was able to provide answers even in the cases in which he could not see his trainer. To give greater credence to his work, the merchant bought two other horses, Muhamed and Zarif, and attempted to instruct them as well. Like Hans, they progressed rapidly—even more rapidly, according to him, because his teaching methods were less confrontational than those of his predecessor.[70] Of all three horses, it seems, Muhamed was the most gifted. After thirteen days, he could already add and subtract. After seventeen days, he could multiply; after twenty, divide. After a month, he was able to answer questions in both German and French, and began to spell within four months of training. Two months later, he could calculate some basic square and cubic roots. Krall even claimed to be able to have short conversations with his horses.

The *Annales* frequently reported on Hans and other animal wonders during this period.[71] The zoological section of the Institut général psychologique even launched an investigation into the possibility of calculating animals (although not much came of it.)[72] On a visit to Elberfeld, Maeterlinck met with Krall and his protégés. He later recalled: "the first shock is rather disturbing, however much one expected it." Krall, according to the author, had not touched his horse and had mostly stood behind it. He had welcomed all restrictions and tests imposed by Maeterlinck. "I assure you that the thing itself is much simpler, and clearer than the suspicions of the armchair critics and that the most distrustful mind would not entertain the faintest idea of fraud in the frank, wholesome atmosphere of the old stable," the author asserted.[73]

Maeterlinck was convinced that the only possible explanation for the impressive phenomenon was telepathy. Krall was providing the answers to his horses through an involuntary transmission of his thoughts. To test his theory, Maeterlinck used his own ignorance in mathematics and asked questions he could not have answered himself. On the first attempt, the horses gave absurd replies, but it was late in the afternoon, the horses were tired, and a new experimenter had performed the test. When Maeterlinck quizzed the horses again the following morning, their answers were correct, even though he did not know the correct solutions himself. Thus, he concluded, telepathy between questioner and horse could not be the major factor to explain the phenomenon.[74] Krall, like von Osten before him, believed that his horses had the necessary intelligence to solve complex cognitive problems. He claimed to have awakened a dormant intelligence in them. For Maeterlinck, however, the theory of animal intelligence was too improbable to be considered alone, and he opted for a mediumistic or subliminal theory. In the end, he concluded, one had to accept that animals could communicate in ways similar to humans, and that, like humans, they could on occasion perceive phenomena outside the range of their normal senses.[75] Their answers to mathematical problems did not depend solely on their brains, but also on the mediumistic abilities they possessed.[76]

Claims of calculating abilities were not limited to horses. In Mannheim, a dog, Rolf, was also discovered to have mathematical aptitudes. In 1913, the SUEP opened its first meeting of the year with a discussion on animal abilities. Edmond Duchâtel, vice-president of the society, wondered if, as argued by Pfungst earlier, the feats could be caused by unconscious movements on the part of the trainer. To him, it seemed improbable that imperceptible motions were always present. More likely, the phenomena of Elberfeld and Mannheim were of the product of a transfusion of the soul in which, for a short moment, the mind of the trainer inhabited that of the animal and controlled its body. If this was in fact the case, it could provide the elusive proof of the existence of the soul as an imponderable fluid.[77] More theories were proposed. After a visit to Rolf in Mannheim, for example, the physician William Mackenzie published an article in the *Annales* suggesting the existence of a double psychism present in every being but more developed in some: the soul-reason, situated in the brain and limited by its capacities, and the soul-intuition, independent of the senses and not attached to any part of the body. Mackenzie wondered if it was possible for the soul-intuition of an individual or an animal to be suggested by the soul-reason of another living being.[78]

Claims of the German animal wonders continued to intrigue the public and feature prominently in the psychical research press. During the Great War, however, news of the once very popular calculating animals from Germany were no longer readily available to the French. Tongue-in-cheek rumors circulated that the horses of Elberfeld had been put in charge of the German cavalry's accounting or that they had died at the front. In 1916, however, readers of *Annales* learned that Krall had sold his horses at the beginning of the war. Muhamed and Zarif were no reduced to wearing saddles and pulling carriages. As for the dog Rolf, he was still at the side of his mistress in Mannheim. "Rolf is patriotic and is rejoicing like a child would at the German victories," she reported. "If this is really the case, cruel disappointments are in store for him, poor Rolf!" the *Annales des sciences psychiques* commented.[79]

PSYCHICAL RESEARCHERS AND THEIR MEDIUMS

if anyone including animals could experience psychical phenomena, mediums were particularly and repeatedly prone to them. Richet distinguished between intellectual phenomena such as automatic writing, telepathy or cryptesthesia, and clairvoyance, which could be found in most of the population, at least to a weaker degree, and physical phenomena such as telekinesis and materializations, which were rarer abilities.[80] As such, mediums were seen as individuals who brought attention to previously unknown human faculties present in all, but stronger and more developed in a few. Maxwell believed that mediumistic abilities were not rare, but were easier to find in nervous individuals. Hysterical tendencies, however, seemed to hinder the development of mediumistic faculties. "I say *instability*, but not lack of equilibrium," he warned. Mediums were not sick, but they were more common among the impressionable, the susceptible, and the moody. Maxwell even believed that mediums constituted the next step in the evolution of humanity.[81] They generally had a vivacious intelligence, attentiveness, energy, and artistic sensibilities.[82] Not every psychical researcher (or metapsychist) agreed with this. Sudre claimed that somnambulists, hysterics, hypnotics, and in general subjects suffering from psychosis made for better mediums. Although mediumistic abilities did not depend on age, sex, or intelligence for Sudre, he did believe that they had been more frequent in the past and were still found in greater numbers in less developed populations. Following Lombroso, Sudre claimed that this could be explained by the fact that civilized means of communications had rendered telepathy and clairvoyance useless. As such: "The function would have disappeared with the need."[83]

Because of their ability to produce psychical phenomena, mediums were of particular interest to psychical researchers. At the end of the nineteenth century, mediums and séances, which had dwindled in popularity somewhat since Kardec's death, returned in full force. In fact, Albin Noiris, a contemporary observer, feared that the 1897 epidemic of séances might claim even more victims than the last outbreak, inasmuch as this new wave had appeared in the guise of science.[84] Flammarion, who had broken his ties with the spiritist movement for some years now, contributed to this renewed interest by organizing a few séances at his observatory in 1898. "In truth, I have always been and I remain a spiritualist; but I have ceased to be a medium. I was once the collaborator of Allan Kardec, the pontiff of this school. I have even been the secretary for a few séances at his home. . . . I signed revelations by 'Galileo.' I had to admit later on, however, that I had been fooled by my own imagination," he would confide to a journalist in 1902.[85]

Flammarion invited Eusapia Palladino, the Italian medium who specialized in levitation, mysterious touches, and moving objects, for séances at Juvisy-sur-Orge, near Paris, where his observatory was located. Eusapia was, by most accounts, a very impressive medium, but she was not above fraud. In 1895, while in Cambridge, England, Eusapia had attempted to produce phenomena through fraudulent means and had been discovered. Like most mediums when caught, Eusapia had not denied it; rather, she claimed that spirits had pushed her to it, that her deceptions had been unconscious and controlled by outside entities. Spirits could be lazy, and it was much easier for them to have her produce the phenomena herself. It was thus the role of the controllers, she argued, to restrain her and prevent the occurrence of fraud.[86]

Eusapia's popularity as a medium had not suffered too much from accusations and confessions of fraud. Despite her reputation as an occasional impostor, she could apparently produce remarkable phenomena even when carefully monitored. After having witnessed Eusapia performing at séances at Montfort-l'Amaury (Île-de-France) in 1897, one observer remarked:

> I used to doubt. Now that I have heard these incredible knocks on the table, have felt this hand pull me back so violently to a place in the room that no one occupied, and have seen this hand with its five fingers spread apart emerge so clearly in front of me between my eyes and the lamp in a place where it was materially impossible for the medium to be, however, and I cite only three main points that have particularly struck me, how can I still doubt the even more extraordinary phenomena reported or written about by credible individuals?[87]

A year later, the Juvisy séances brought Eusapia back to France and Flammarion back with full force to a world he had been neglecting for thirty years. The event was publicized. Eusapia was not a cheap medium to invite, and in order to pay for her services, Flammarion invited journalists, writers, and scientists in exchange for their financial contributions.[88] Those in attendance included Flammarion and his wife, Sylvie; the writer Alexandre Bisson and his wife, Juliette; the journalist Jules Blois; Charles Richet; and Eugène Antoniadi, Flammarion's assistant. With the exception of Antoniadi, this distinguished group of participants had observed numerous other séances before attending the Juvisy séances of 1898, and their impressions would be given great credence.

For most of those present, the séances were a success. For Antoniadi, however, the results were disappointing. In a manuscript report, he was critical of both the medium and the participants. The atmosphere of the proceedings had rendered the occurrence of any phenomenon suspicious. "The majority of these people [the participants], being individuals already converted to 'spiritism,' were not, of course, well prepared to observe strict vigilance over how the 'phenomena' were produced," he observed.[89] For the most part, the séances had been boring, the phenomena had taken a long time to manifest themselves, and Eusapia had encouraged the participants to talk and sing while waiting, presumably to help her produce the phenomena. For Antoniadi, these noises, diversions, and distractions could have easily been used to cover up fraudulent attempts: "Mme Fourton has, in addition, literally deafened us with her Wagnerian songs and others. . . . assembled as we were for a serious scientific investigation, where silence was called for, these noisy manifestations were most ridiculous."[90]

At a séance, participants had to abide by the medium's demand if they wished for results. After all, mediums always claimed that the sympathy of their audience was vital to their success and often explained their failure by referring to a negative ambiance or a specific participant. They could ask for specific changes in the proceedings, often leading to a reduced possibility of control. At Juvisy, Eusapia had asked the openly skeptical Antoniadi to move farther away from her.[91] She had refused to be searched prior to the séances. She had asked to be seated on a kitchen chair rather than a more comfortable but more restricting armchair. Every phenomenon witnessed had taken place within her reach. When she had provoked objects to move, it had been on the table in front of her, while making the kind of mysterious hand gestures that could easily hide a hair used to move the object.[92]

In his report, the astronomer's assistant expressed his exasperation at the lack of seriousness with which the séances had been conducted. In his words, the mystery and the spectacular quality of the proceedings had been turned to trivialities: "The 'medium' has pretended to enter in trances by yawning and feigning the hiccup, while her face was taking a malicious expression, almost demonic," he described.[93] The phenomena, when they had occurred, had been ridiculous and clearly fraudulent, he concluded: "There is only fraud, and fraud only. None of the phenomena are authentic."[94] If Eusapia had not been caught, it had been because the controllers had been eager for something to happen and had failed to control the medium adequately. It was true that honorable and respected individuals had been present, but Antoniadi was shocked by their complete lack of discernment: "we saw M. Bisson, editor of the *Annales politiques et littéraires*, seriously raise the question of 'whether the book would not have dematerialized itself while crossing the curtain to rematerialize itself after the crossing'! It is unfortunate for humanity to note that it is men of this kind who, most of the time, attain the most elevated of situations and the greatest honors."[95] As for Richet, Antoniadi recalled:

> Although at this moment Dr. Richet found himself six meters away from Eusapia, at the far end of the living room, and that, consequently, he had not seen how the "phenomenon" had been produced, he nonetheless shouted: "Bene! Bene! What a beautiful phenomenon! What a beautiful phenomenon!" These exclamations depict this man in his true colors; they disqualify him forever from expressing any scientific opinion on these matters.[96]

While Antoniadi had not been impressed by the discerning judgment of the participants, men like Richet and Flammarion were reputed to be skeptics in the spiritist circles. Writing on the Montfort-l'Amaury séances a year earlier, the photographer Guillaume Fontenay had praised Flammarion's abilities to judge mediumistic phenomena.[97] By then, Fontenay, a respected participant at séances himself, was making a significant contribution to psychical research with his work on photography. Photography was a powerful means of persuasion for both mediums and psychical researchers, but it was not above fraudulent practices.[98] Spiritists had made ample use of it to record manifestations of the spirits and sometimes the spirits themselves. In 1875, Pierre-Gaëtan Leymarie, then editor of the *Revue spirite*, and Édouard Buguet had been put on trial for fraud, convicted, and sentenced to one year in prison. It seems that they had deceived spiritists with fraudulent pictures of ghosts, taken by Buguet and published by Leymarie. Buguet's studio had been raided and dummies and photographs of heads on cardboard had been found. Faced with the evidence, Buguet had confessed, but later retracted, claiming that it was only his assistants who had used the dummies while he was away from the studio. He had insisted that two-thirds of the ghost photographs were authentic.[99] In his work, Fontenay focused on different ways to evaluate photographs and to prevent such fraud. Photography, for him, was a tool both to control the medium and to push psychical research further. First, it could work to verify what our eyes saw, serving as a safety measure or control; and second, it could be used to record phenomena that were not visible to the naked eye, a photography of discovery or research.[100] The use of photography in psychical research understandably provoked great interest. Its promise even led to a contest

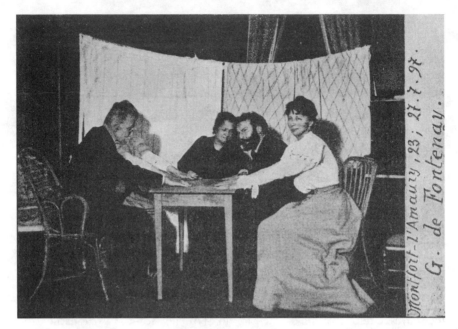

One of the séances held at Montfort-l'Amaury, near Paris, where Guillaume de
Fontenay was so impressed with Camille Flammarion's ability to judge a medium.
In the photo, Flammarion and Eusapia are leaning toward each other, with three
members of the Blech family in the foreground. From Guillaume de Fontenay, À
propos d'Eusapia Paladino. Les séances de Montfort-l'Amaury (25–28 juillet 1897).
Compte rendu, photographies, témoignages et commentaires (Paris: Société d'éditions
scientifiques, 1898), fig. 8.

in 1908 when a Société d'étude de la photographie transcendantale, discussed on
a few occasions in the *Annales des sciences psychiques*, opened a year-long competition in which two prizes, of three hundred and six hundred francs respectively, were to be given to the individuals who submitted photographs of invisible
beings or unknown radiations.[101]

Since fraud was frequently uncovered at séances, control was essential for those
interested in formulating scientific explanations of the phenomena witnessed.
Authenticity could be confirmed only once the participants had eliminated the
possibility of conscious or unconscious fraud by the medium. The setting and
the proceedings of séance, however, made any attempt to scrutinize the phenomena difficult. The medium, usually a woman, always reserved the right to control
the proceedings. If they wanted to witness psychical phenomena, observers found
themselves at the mercy of her whims. She usually asked for darkness and might

encourage her audience to talk, even sing. One author even talked of "the 10 commandments" of every good medium: obscurity, dark cabinet, all hands occupied in a chain, trance, noise and conversation, no photography without permission, variance in the medium's power to be expected, understanding required from everyone in the audience, long wait, and excessive sensitivity, leading to the interdiction against touching materializations and ectoplasms.[102] If it was true, as mediums claimed, that the phenomena were very sensitive and could only be reproduced in specific environmental conditions, it was an unfortunate consequence for psychical researchers that those conditions were also conducive to fraud. The séance, by its nature, encouraged fraud in mediums, rendered control difficult for psychical researchers, and made it hard for the scientific world to accept the reality of the phenomena.

There is no denying that deception was a significant part of séances. If, for some, it was the rule more than the exception, the revelation of a fraud did not have dramatic consequences. Everyone suspected Eusapia of having attempted fraud on numerous occasions. The psychologist William James even wrote: "It is a known fact that Eusapia cheats by every means at her disposal when she is allowed to do so."[103] Yet she continued to have a successful and lucrative career. Accusations of fraud were not too damaging for psychical researchers either. For example, a few years after the Juvisy séances, many criticized Richet for his support of the Villa Carmen materializations. Starting in 1902, Carmencita Noël published reports of séances held at her home in Algiers, where spectacular phenomena were occurring. Marthe Béraud was materializing the spirit of one Bien-Boâ and his sister Bergolia for the visitors of the Villa Carmen. Béraud had been engaged to Noël's son before he had died in Congo. In 1905, Richet traveled to Algiers accompanied by spiritist Gabriel Delanne to witness the Noël séances. He saw the ghosts produced by Béraud, even touched them, and declared the phenomena genuine.[104]

Evidence of fraud began to surface in 1907, however, when a physician from Lisbon claimed to have obtained a confession from Béraud herself. In a report for the XVth International Congress of Medicine, Dr. Rouby exposed the fraud revealed by the staff of the Villa Carmen. For example, a certain M. Portal had confessed to helping out with séances when he could. He recalled: "One day I began to talk Provençal aloud: the ecstatic general announced that Bien-Boâ talked Hindu! . . . What memories, my God! I could talk of this for hours! How is it possible that afterward we then took this seriously!"[105] Another member of the staff, Areski, had admitted to having participated in the fraud, often helping Béraud and impersonating Bien-Boâ during séances. He had even claimed that it had been Noël herself who had taught him how to move in his disguise, saying she knew it was fraud but also believed it at the same time.[106] On Richet and Delanne's visit, Areski confided: "Marthe has told us everything. First of all, she was a bit scared, not of M. Delanne, but of the other [Richet], who is, apparently, a great scholar. She did not want to do Bien-Boâ in front of him, but when she saw that he was not shrewd at all, she did not worry about it."[107] Béraud herself and her father also appeared to have confirmed the fraud.[108] Roupy then went on to explain that it had been decided with the Bérauds that the fraud would be revealed to Richet. According to him, a letter explaining the events was sent to Richet by the father, a letter that the scientist had failed to publish.[109]

Claims of fraud did not stop Béraud from participating in other séances. In 1909, under the name of Eva Carrière (often shortened to Eva C.), she began a new career as a medium, denying any fraud at the Villa Carmen and distancing herself from her past mediumistic productions. Eva C. went on to become a renowned and successful medium. As for Richet, he continued to maintain the authenticity of the phenomena he had witnessed at the Villa Carmen even years later. On Béraud's supposed confession, he even wrote that if this was in fact true, more than anything else, the statement revealed the medium's mental instability and should not be taken seriously.[110]

This was not the only time Richet was accused of extreme gullibility. Pickmann, a telepathic medium and later on magician who had arrived in France from his native Belgium in 1892, had been observed by a few scientists including Richet. Writing later of his encounter with the medium in his *Traité de métapsychique*, Richet recalled a particular experiment: "I did with him an experiment that is, I believe, irreproachable where method is concerned." Richet had chosen a single card out of a deck of fifty-two without Pickmann present. He had then looked at it attentively and tried to represent it to himself visually. Finally, he had called in Pickmann, who all of this time had been in the other room, and asked him to determine the chosen card, while he, himself, remained with his back turned. "He had complete success ($\frac{1}{52}$) in the first experiment, success that hugely surprised and delighted us both. But the later experiments were not successful (three failures)."[111] In 1924, Pickmann was interviewed by the journalist Paul Heuzé and confessed to a number of deceptions, including some with Richet. Of this particular session, he recalled: "[I]magine that, finding myself in the great antechamber, I was providentially able to see the professor choosing the card through the half-open door,: I remember it; it was the nine of clubs. But you can well imagine that afterward, when Richet, now, I think, suspicious, went about it in such a way that I could no longer see anything really, I could not 'guess' any card. Of course! Such an experiment is impossible without a trick."[112]

Richet was not the only scientist Pickmann proudly claimed to have fooled. He boasted to Heuzé about how easily he had been able to convince Cesare Lombroso, with whom he had chiefly collaborated, of his authenticity:

Ah! This beats everything! I have never in my entire career seen such a sucker! Any prankster could tell him some stories; quickly he took it down, and bang! Observation number 4613! . . . With me, he came to my hotel every day; it's plain to see, incidentally, the publicity this brought me! And he brought complicated apparatus of all kinds that he would put on my body: I have never been able to understand what he was looking for! . . . As for my *fluid*, here is what I can tell you about it. He annoyed me so much that one day I told him: "Master, I'll try my force on you: turn around and I'll prove the attraction of the shoulder blades." I placed myself behind him, calmly caught hold of his robe, pinched it between my thumb and the index finger, and pulled. I give you my word, Monsieur Paul Heuzé, that that's what I did. Lombroso didn't notice a thing, and he was delighted. "All the same, you've got some nerve!" my secretary said after he had left. "I wanted to show you how far the foolishness of a great scholar can go," I told him.[113]

Of course, not all mediums were sincere. In the end, however, accusations or even confessions of frauds were never insurmountable or too damaging; in fact, they seemed to fuel interest in psychical research and create new challenges and puzzles for adherents to work out. Mediums could always deny earlier confessions; they even could admit to a fraud caused by the influence of the spirits. At séances, psychical researchers had little choice but to go along with their demands and monitor them as closely as possible. If they were developing a science that was riddled with fraud, there was little they could do about it. Psychical research, by definition, respected the experiences and claims of its subjects. If one believed in its possibilities, one had to accept the problems and the limits associated with the field.

◦◦

Psychical research developed in the context of an increased professionalization of the sciences. By the end of the nineteenth century, the separation between scientists and lay public had been established. In contrast, the boundaries of psychical research remained fluid. Many psychical researchers admitted to having few or no scientific credentials. Contributions by both men and women without scientific backgrounds were welcomed in the field. Books on the topic were written with a large audience in mind. Research results were presented with photographs and made accessible and often entertaining. Psychical researchers built their field in societies and journals open to all. Everyone was encouraged to subscribe, to send in recollections of their own encounters with the supernatural. The lines between reader, subject, observer, and experimenter were never rigid; neither was a distinction made as to which space could be used to practice psychical research. Observations could be done everywhere: at home, in private or public séances, in open society meetings, even at the music hall.

In the first few years of the *Annales*, there was great optimism on the part of its founder as to the possibilities of this unrestricted and open approach, but by the beginning of the Great War, the journal was loosing its momentum. Only a few issues of the *Annales* were published during the war, and, by 1919, it had petered out. Interest in the seemingly supernatural had not died down, but it had changed. The carnage of the war and the experience of death meant that many séance participants turned inward and focused on the consolatory side of the practice. Spiritism continued to exist, but its claims to a scientific dimension were largely abandoned. Moreover, both the subject matter and the methods of psychical research in France were changing. The up-and-coming generation of psychical researchers favored a more restrictive, less popular approach. Gone were the public appeals and the easy accessibility. Starting in 1919, the Institut métapsychique international would propose a new approach to the investigation of the supernatural in France.

The Rise and Fall
of Metapsychics

In October 1918, Dr. Rocco Santoliquido, a public health official, informed numerous friends and colleagues of his intention to create an international institute of metapsychics in Paris. Far from eclipsing national and regional societies, he wrote, the new institution would allow better communication and collaboration across the discipline.[1] Santoliquido had obtained funding for his institute from the French industrialist Jean Meyer, a committed spiritist, who had proposed to finance both an institute of metapsychics and a new spiritist society. While the latter would propagate Kardec's doctrine, Santoliquido's Institut métapsychique international (IMI) would be dedicated to the scientific study of spiritist and other related phenomena, recognizing their authenticity, but not restricting itself to the promotion of spiritism or any other doctrine.

The institute would be focused on the study of *métapsychique* rather than *sciences psychiques*. Until then, *sciences psychiques*, inspired by the English "psychical research," had been preferred for describing the field in France. The term *métapsychique*, created by Richet in 1905 to designate "a science that has as its subject mechanical or psychological phenomena caused by forces that seem intelligent or by unknown latent powers of the human mind," had largely failed to attract a following in the psychical research community so far.[2] Thus, it was clear to all concerned that the choice of this name implied a new orientation and an attempt to break with a past tainted with embarrassing accusations of frauds and lack of scientific rigor. There was no doubt that the IMI would be in a position to dominate psychical research. Its financial assets alone meant that it would play a leading role in the field, both nationally and internationally, and the founder's inclusion of the word *internationale* in its title clearly expressed determination to play an important role abroad. Santoliquido's efforts to establish such an institute in Paris could only be interpreted as a bid to increase the French influence in the field.

For its members, the IMI represented the promising future of metapsychics, psychical research, or any other serious work on séances. Throughout the 1920s, metapsychists participated in international congresses of psychical research, trying to impose their methods on others. Unlike psychical research, which had produced very few interesting results in the years since its inception, they claimed that metapsychics would finally provide the field with the scientific credibility it so wished to obtain. Of course, this systematic attempt by IMI members to impose themselves on a very dispersed and diverse group would not always be met with approval, and tensions became apparent both inside and outside the institute throughout the 1920s and early 1930s.

THE EARLY YEARS OF THE
INSTITUT MÉTAPSYCHIQUE INTERNATIONAL

In 1905, the physician and enthusiastic spiritist Gustave Geley wrote a pamphlet denouncing psychical research as it had been practiced until then. In his assessment, for more than half a century the field had not witnessed serious progress in its observational and experimental methods. It had simply plodded along; the same empiricism, the same attempts, the same uncertainties, and the same obstacles remained. Despite the considerable efforts that had been made, little progress had resulted; and, above all, no certainty had been obtained. The public remained, for the most part, indifferent to the research, and most scientists were hostile to it. But how could it be otherwise? For Geley, the incessant presence of mysticism in and around psychical research, the pessimism of researchers in the face of dubious results and constant ridicule, and the movement's lack of cohesion guaranteed its failure.

But all was not lost, Geley believed. Psychical research could be saved provided a strong unifying body or institution was created to combat public indifference and the hostility of the scientific community. In France, psychical research societies had adopted a passive approach that would never produce interesting results. Most societies avoided theorizing and limited themselves to straightforward presentation of observations at rarely held meetings. There was evidently a need for an active modern society, one that would not restrict its activities to collecting documents and investigating occurrences brought to its attention, but would promote experiment, research, and analysis. In Geley's opinion, this was "the only efficient way to triumph against the inherent differences in psychical studies."[3]

Geley's plan for an international institute promoting psychical research included a well-equipped laboratory, an information office where relevant documents would be collected and kept for consultation, and a bulletin promoting members' activities and findings. This institute would provide education for mediums on matters of control and fraud. It would encourage collaboration between nations and possess the resources necessary to organize the study of psychical phenomena in non-European countries where mediumistic abilities had often been reported as more prevalent. It would carry out extensive projects such as the collection of sealed envelopes from various individuals containing photos, physical characteristics, psychological profiles, writing samples, signature, fingerprints, and footprints, thus building a database that could be used to authenticate later communications with the dead.[4] It would mark the beginning of the serious study of survival after death. The project required funds and determined researchers willing to adhere to the scientific method and devote their time and effort to the cause. Such an enterprise, Geley pleaded, would be worth every sacrifice.[5]

It was not until he met Santoliquido that Geley's project became a reality. A physician by training, Santoliquido had been working on public health issues, particularly those related to epidemiology, in his native Italy since the 1880s. In 1907, convinced that only a coordinated effort at the international level could efficiently combat the spread of epidemics, Santoliquido had accepted the presidency of the Office international d'hygiène publique, an institution created to administer international sanitary conventions and coordinate the circulation of epidemiologic information.[6] He moved to Paris and held the post until 1916. Santoliquido arrived in the capital already interested in spiritist manifestations. His first encounter with the phenomena had occurred only a few months earlier, when he had found his son, his niece, and some of their friends performing a séance at his home. At first he mocked such practices as superstition, but the group prevailed upon Santoliquido to put the table to the test, and he was startled when the presumed spirit correctly answered the seven questions he had thought of but had not spoken aloud. Santoliquido was converted and began conducting his own experiments in the field. He concluded, like many before him, that one of the participants (in this case his niece, Louise) had been unconsciously causing the phenomena. He also became convinced of the reality of what he called supranormal knowledge, the belief that humans are not limited by their five senses but can obtain information by other means. His interest in séances did not, however, become a serious preoccupation until the Great War, when he met Geley and hired him as his personal secretary. The two men began to talk about metapsychics and the need for a serious international laboratory devoted to the development of this science, and Geley—who had not forgotten his earlier dream of an institute—encouraged and helped Santoliquido to realize the project.[7]

Through fortuitous circumstances, they met the industrialist and spiritist Jean Meyer who agreed to fund both a spiritist institution, the Maison des esprits on the rue Copernic, and a scientific laboratory, the IMI, on the avenue Suffren. From this initial laboratory established in 1917, and with the financial support of Meyer, Santoliquido and Geley were soon able to build the institute.[8] All they now needed was the moral and material support of the psychical community. "The successes of our armies, [and] the triumph of the ideas that are dear to us, allow us to foresee humanity's imminent entrance into a new, superior phase," Santoliquido wrote to some of the most respected and established psychical researchers in October 1918. "Along with definitive peace among men, victory must bring something more: a glimmer of hope to those who mourn their children killed for an ideal, a ray of truth to researchers of honesty."[9]

While some received the news of this project with enthusiasm, not everyone shared the founders' enthusiasm for creation of the institute. On October 16, 1918, Sir Oliver Lodge, a respected physicist in the field of electromagnetics, an enthusiastic psychical researcher, and one of the most prominent members of the Society for Psychical Research (SPR), confided his reservations about Santoliquido's project in a letter to Richet: "There can be no doubts about the importance of the research and the work that will be pursued in this way in our countries once peace has liberated energies. The question is whether the moment has come for unified international action. It is on this point that I am in doubt."[10] For Lodge, the national traditions of psychical research were still too strong to allow for the creation of an international institute. Whereas the English-speaking world had so far mostly focused on the question of the survival of the soul, the French had been less inclined toward this particular line of research and had spent more effort on the medical and psychophysical aspects of psychical research. Expressing more than a simple nationalistic objection to the project, however, Lodge also feared the potential dangers such an institution would carry with it. A stronger organization implied a greater visibility. For the English physicist, the discipline was not yet ready for the kind of scrutiny such an institute would bring from various scientific and popular directions.[11]

In spite of such apprehensions, by the early spring of 1919, the IMI was ready to be launched. On 11 April, 1919, a preliminary directing committee held its first meeting. The financier and spiritist Meyer welcomed the participants, and Geley presided over the proceedings.[12] In June 1919, the IMI held its first official meeting with its new committee consisting of ten members, some of whom, such as Richet, Gabriel Delanne, and Camille Flammarion, were well known in spiritist and psychical research circles. Over the next ten years, the central committee membership would not change much, but by 1930, it had grown in size and become more international in character (now including such members of the psychical research community as Oliver Lodge, Hans Driesch, and Ernest Bozzano).[13]

In the first few years of its existence, the IMI flourished. Its laboratory rapidly moved to a more spacious location on the avenue Niel. There, the institute occupied two floors until 1955. The first floor housed the laboratory, which included a mechanical workshop, some photographic equipment, an electric scale, and all the necessary apparatus for basic chemical experiments. All around the room, electric lamps with controllable luminosity were placed. Red lights of various intensities were also available. The second floor housed a library and a conference room.[14] In keeping with its mission to educate the public and provide metapsychists with a congenial setting in which to interact and collaborate, the institute organized public lectures on subjects ranging from the origins of metapsychics and the history of the phenomena to more focused topics and even occasional presentations by invited mediums; it organized public séances with mediums collaborating with the institute and later held weekly public consultations on metapsychical issues.[15]

In October 1920, the IMI started its own journal, the *Revue métapsychique*, which appeared every two months between 1920 and 1940. During the Second World War, its activities slowed down significantly (only one issue of the journal was published at the time) but, in 1946, publication resumed and continued until 1982. In its early years, the journal published research on mediumistic productions and other faculties of the mind, extreme manifestations of faith and mysticism, and phenomena of a similar kind produced in the East. In addition, it provided discussions on ways of reconciling metapsychical phenomena with the laws of physics and biology and presented practical considerations on the role of metapsychics in medicine and the justice system. It also invariably contained book reviews and news of the field around the world. Over the years, the *Revue métapsychique* continued to be presented as a serious and unifying venture into the scientific study of the psychic.

Throughout all of this, the director of the institute played a significant role in shaping its research agenda. The early years of the IMI under Geley's direction were marked by his own interest in ectoplasm—the mysterious substance claimed to emerge out of a medium's body and develop into arms, legs, heads, or even entire human bodies or objects. A physician by training, Geley had developed an interest in the occult after observing many occurrences of clairvoyance, somnambulism, and premonition in his medical practice. Although he remained more sympathetic to spiritism than others at the institute, he always kept his distance from the more spiritual aspects of the field and preferred to focus on the experimental and theoretical side of the phenomena. While he believed in the immortality of the soul, reincarnation, and communication with the dead, he saw them as rational truths rather than revealed ones.[16] Geley preferred the most spectacular of mediumistic productions, and, as director of the IMI, he recommended that metapsychists concern themselves with the physical productions of séances instead of more intellectual phenomena such as automatic writing, for which fraud could not be as easily detected.

Geley was an accomplished researcher. Having first witnessed numerous experiments with the medium Eva C. and her protector Juliette Bisson in the 1910s, he had begun to pursue his own observations into ectoplasmic productions in the 1920s with the Polish mediums Franck Kluski and Stephan Ossowiecki. He included most of his experiments and observations on mediums and presented a comprehensive discussion on the particular nature of such scientific work in his last book, *L'ectoplasmie et la clairvoyance* (1924). Metapsychics, he explained, was the science of mediums, individuals in whom a hereditary tendency of decentralization of the mind had been reinforced by the practice of mediumship to produce exceptional and abnormal abilities. It was not uncommon to explain mediumistic abilities as the consequence of a splitting of the mind. This was not a disease or a pathology, but a superior ability exhibited by some gifted individuals.[17]

Mediumistic phenomena were fragile and difficult to provoke, however, Geley often reminded others. They depended upon an array of conditions: very low light (too much light would disturb the medium's trance and impede the materialization process), the good health and positive mood of the medium on the particular day of the experiment; and his or her confidence and ease with the observers. The mood of a séance was everything because each participant, each observer, became part of the experiment. In a friendly atmosphere, the medium's faculties would be reinforced and multiplied. The ideal number of participants was from four to seven. Patience and unity were essential. A young and healthy audience was conducive to success; so was one that was informed and ready. Observers had to let the phenomena come without trying to provoke them and know how to control them when they did come. According to Geley, to blame a medium for a failed séance or to congratulate one on a successful production was always misguided: "*merit and responsibility are always collective*, as are the experiments them-selves," he wrote.[18]

Geley remained the dedicated director of the institute until his death in July 1924 when, returning from Poland, where he had gone to witness mediumistic productions, his plane crashed near Warsaw. Under his directorship, the institute had flourished, and he had contributed an article to almost every issue of the *Revue métapsychique* since it first started. "Geley was the soul of the great scientific movement of metapsychics," Richet wrote in the late summer of 1924.[19] Some even believed that Geley's dedication to proving the survival of the soul continued after his own death. Shortly after the plane crash, his friend Marja Wodzinska organized a séance in Warsaw hoping that the dead metapsychist would come to visit her. She later revealed that three days after his accident Geley had come to her in a dream and said: "Tell everyone that I am now completely happy." The séances began once Geley's widow arrived from Paris for her husband's funeral. Jean Guzik, a medium who had worked closely with Geley in 1922 and 1923, at-tempted contact. Initially, nothing happened, but once Geley's widow had returned home, the director appeared in three successful séances. During a first séance, an apparition who cordially embraced participants and spoke French with some con-fusion was identified—with some reservation—as the French metapsychist. At the second séance, a severed head believed to be Geley's after the plane crash appeared to the participants. Finally, a full body was said to have materialized during the third séance. When Wodzinska expressed her disappointment that she did not have paper and pen for the apparition to write a few words, it said: "I can-not do anything" and with this disappeared, leaving no physical evidence of his visit or means to confirm his survival. In December 1924, Wodzinska reported that she was still waiting for another visit from her friend.[20]

Eugène Osty, also a physician, was chosen to replace Geley as director, a posi-tion he held until his own death in 1938. Unlike his predecessor, Osty was never a spiritist, not even a skeptical one. Under his lead, the IMI moved further away from explanations of séances involving spirits of the dead. For the most part, it also neglected studies of ectoplasmic productions to focus more energy on Osty's own interest in supranormal knowledge. Osty had been interested in metapsychical phenomena since the 1910s. In 1922, he had published a book titled *La connais-sance supra-normale*, in which he had differentiated between two kinds of knowl-edge, normal and supranormal knowledge, the latter obtained through "the char-acteristics that some people have to learn immediately, without the use of known senses (or rather without the known use of known senses) or of reason, the moral, intellectual, organic, and social characteristics of human individualities put in contact with it, and the past and future course of their lives."[21] For Osty, either the brain possessed undiscovered physiological properties or it was not the only site producing human thought.[22] In 1932, he published *Les pouvoirs inconnus de l'esprit sur la matière* with his son, the engineer Marcel Osty. The book discussed experiments with the Austrian medium Rudi Schneider in which father and son had observed the partial absorption by infrared light of an invisible substance projected during telekinesis.[23]

Wishing to provide the institute with a more serious and scientific agenda, Osty rapidly decided to reassess the IMI's initial aims and progress. In 1925, members of the directing committee met to discuss his new program of research. After five years of its existence, the time had come to reassess the mission of the institute, Osty contended. First, the IMI should now seek to demonstrate the reality of metapsychics and supranormal knowledge in humans to the scientific elite by providing clear and simple observations of the various phenomena; and second, it should promote experimentation and research on the psychological, physiological, and physical aspects of such phenomena, which would lead to the greatest discovery regarding humankind.[24] Much in the way the study of hospital patients had allowed psychiatrists to discover the subconscious, the study of gifted individuals would reveal the existence of supranormal knowledge and abilities over matter. Unfortunately, the institute lacked the financial resources to pursue this goal completely.[25]

On November 16, 1925, the committee approved the report. It expressed its gratitude to Meyer—who had hitherto been the institute's sole financial backer—for his generosity, but it was agreed that it was now time to look for other sources of funding. It was decided that, starting in 1926, the IMI would adopt a more aggressive financial plan and strongly encourage other wealthy and cultivated individuals to contribute.[26]

Beyond the fact that the institute was funded solely by an individual lurked a greater problem. Under Osty's leadership, the IMI's dependence on the financial support of a convinced spiritist became a source of tension. As an institute wishing to legitimize a field of research that was still in large part the purview of nonscientist enthusiasts, the IMI was in a delicate position. This was not a new problem. Psychical researchers and metapsychists were always faced with the fact that some of them, and most of their audience, were not scientists. The fact that the very survival of the field depended on good relations with potential financial supporters, often spiritists interested in lending greater credibility to their beliefs, while scientific legitimacy, if at all possible, required a clear separation from the spiritist doctrine, left psychical research and metapsychics in an impossible situation.[27]

Two incidents in particular illustrate tensions that the association with Meyer produced under Osty. In 1926, René Sudre, then a member of the IMI and an important collaborator at the *Revue métapsychique*, was informed that his contribution to the journal was no longer required. In September of that year, Sudre wrote to another member of the institute of his surprise upon learning the news and his suspicions on the motive behind such a decision. "[N]o one will mistake the meaning of this measure," he declared. "In staying in an organization financed by a spiritist, I cause too many problems for the spiritist cause. That is great clumsiness, of course, and it will discredit the institute from a scientific point of view—not because of my modest personality but by the confessional character henceforth to be attributed to the house on the avenue Niel."[28]

Sudre was never a spiritist. In contrast to Geley or Richet, he did not give much weight to the hypothesis that metapsychical phenomena might enlighten scientists on the survival of the soul and the existence of an afterlife. Moreover, his portrayal of those who did consider the spiritist or occultist hypotheses had always been derogatory. The same year that he was pushed aside by the IMI, Sudre had published a book titled *Introduction à la métapsychique humaine*, in which he had made no secret of his conviction that the phenomena were natural rather than supernatural in origin.[29]

The tensions brought on by Sudre's dismissal led to the resignation of a second member of the IMI. On December 12, 1926, Daniel Berthelot, a member of the Académie des sciences and the IMI's committee since the early 1920s, left the institution. In his letter of resignation, Berthelot informed the president that, having learned two days earlier of Sudre's dismissal and the reasons for it, he now wished to distance himself from the institute.[30] Santoliquido's reply came three days later in a letter in which he expressed his surprise at a resignation based on such tenuous motives and on the "sad intrigues" of Sudre.[31]

In 1928, trouble arose again, this time from Jean Meyer himself. Feeling that he was getting old, Meyer made provisions to ensure the financial support of the institute after his death. In July 1928, he created the Société des études métapsychiques. In his will, he gave eight shares of the society to the IMI, eighty-six shares to various individuals, and sixty shares to his butler, a man named Forrestier.[32] Unfortunately, this left the institute with little control over its own finances once the spiritist was gone, a situation that made Santoliquido very nervous. The tensions that had remained subdued for a decade now erupted. In January 1929, Santoliquido wrote to Meyer that he feared that the IMI was being forced in the direction of spiritism. Conflicts between the spiritists and the institute were not new, Santoliquido recalled, citing a previous incident in which Meyer had warned Geley that the IMI was not in a financial position that allowed freedom of research:

> At the time of the provisional laboratory on the avenue Suffren, and from the first years of the Institute on the avenue Niel, Geley often confirmed to me that he was saddened by your insistence that he give to the direction of the IMI a "bit of help" (his words) in the spiritist direction. Before leaving for Warsaw, in 1924, where he met his death, he told me that he had implored you to destroy the contract (contract that assured his means of existence) in which you had given this orientation to his direction. Fearing the unfortunate scandal in which this document might one day involve his work, he insistently asked me to help him get you to destroy it.[33]

The conflict continued. A few weeks later, Santoliquido warned Meyer:

> Do not suppose that we can continue to treat the Institut métapsychique as an organization committed to genuine, free science if we discover that it is forever at the mercy of a will other than that of its directing committee. For, [and] I cannot say it often enough, a founder is one thing; his successor is another. . . . such incidents have little importance when they concern the founder. The feelings of gratitude that we owe him excuse everything. With a successor holding the Institut métapsychique at his mercy, such a situation would not be possible.[34]

In a letter to Osty, Joseph Maxwell, then a member of the IMI's committee, presented Meyer's side in this affair:

> M. Meyer assures me that he has given total liberty to the institute, and [asserts that he] has the right to specify the direction scientific research should take. After all, he founded an Institut métapsychique, and the status of his foundation was obviously under his control. He has defined its objective, which is the search for proofs of survival, but he has specified that this search should be accomplished scientifically. This specification was made to avoid meeting the same fate as the institute founded by his friend YOURIEVITCH [the IGP], which has become a banal psychological institute busying itself with the psychology of bees, ants, and termites. This example justifies his precaution.[35]

Maxwell's plea came too late. By then, the breach between Santoliquido and Meyer had gone too far for reconciliation. On March 15, 1929, Santoliquido resigned his position as president of the IMI. "[H]e had no illusions about the sad future of the IMI. He knew what to expect from the person [Forrestier] entrusted with the fate of the institute," Osty later recalled.[36]

Charles Richet replaced Santoliquido as president.[37] In May 1929, Meyer transferred ownership of the avenue Niel building to the IMI and announced that he would endow the institute with 12,500 francs per month.[38] The situation improved, but it was only a temporary truce. After Meyer's death in 1931, frictions arose once again. Like Santoliquido, Meyer's family was suspicious of Forrestier, who had risen rapidly from being a simple butler to become Meyer's secretary and finally his personal medium. In the last few years of his life, the financier of the IMI appeared to have been under the spell of this man. After Meyer's death, his heirs contested the will, arguing that too much had been given to Forrestier. In the conflict that followed, the IMI supported the annulment of the will, leading Forrestier to cut the institute's funding.[39] Meyer's death and the clash with his successor thus marked the end of financial stability for the institute. The IMI survived, but only with constant worry about its monetary situation.

TOWARD SCIENTIFIC LEGITIMACY

The institute's attitude toward "official science," as its members referred to it, was ambivalent. On the one hand, members of the IMI saw themselves as scientists and their discipline as a science; they glorified scientific reasoning and the scientific method. On the other hand, they spoke of their persistent and frustrating struggle to penetrate the sciences, realizing that their field of research had yet to be considered scientific by the wider community. Metapsychists worked toward the inclusion of new forces and powers of the human mind into "official science." They saw themselves at times as scientists and at times as future scientists, but they also attacked science for its narrow-mindedness. They were resentful that their burgeoning discipline was still ignored or pushed aside; that its observed phenomena were, when at all considered, redefined outside the metapsychical set of preferred explanations. Complaints about the general attitude of scientists to metapsychics were frequent. Metapsychists often felt persecuted and ridiculed by a scientific establishment unable to recognize the potential of their work. They saw their cause as a heroic one. Future generations would recognize them as martyrs of science, true courageous soldiers.[40]

Until the 1920s, the psychical research community had interacted little with French scientific institutions. Of course, a number of those interested in psychical phenomena were respected scientists, Richet being the most prominent of them. If Richet's scientific career was unequaled in the metapsychics community, his example suggests, more than anybody else, that metapsychists were not necessarily the outsiders they perceived themselves to be, that they were not always ridiculed or pushed aside by scientific institutions, and that they could sometimes be respectable members of the metapsychic and the scientific communities simultaneously. Richet's involvement with spiritists, psychical societies and, later on, with the IMI were always very public. Upon retiring from the Faculté de médecine in 1925, he even chose metapsychics as the subject of his penultimate lecture.[41] Although his work with mediums was never officially recognized or even appreciated by his colleagues at the university, Richet does not appear to have been ostracized for it.

Richet may have constituted an exception, but there were certainly other, less successful scientists who pursued metapsychical research with equal enthusiasm. If, like the spiritists, the occultists, and the psychical researchers before them, metapsychists talked about "official science" as something they could not penetrate, the reality was not that simple and obvious. Many metapsychists were scientists by training and held scientific appointments. Geley, Maxwell, Osty, and Santoliquido, to name a few, had all held positions in a medical establishment at some point during their career. Moreover, metapsychical works were not completely ignored by other scientists. For example, Richet's 1922 *Traité de métapsychique* was presented at the Académie des sciences and reviewed by Pierre Janet in the *Revue philosophique*.[42] If it remains true that there was little place made for metapsychical concerns in academies and universities, metapsychists were never completely ignored nor were they persecuted to the extent that they described. There was certainly a reluctance to discuss the field seriously in academies and universities, but never an outright refusal.

One clear example of the willingness to consider this type of work emerged in the early 1910s when the Académie des Sciences instituted the Fanny Emden Prize. In 1911, the academy introduced a new prize to be given every two years for a book on psychical research. Discussion on the creation of this prize had begun in 1904, when two members of a prominent banking family, Fanny and Juliette de Reinach, mother and daughter, had expressed the desire that a prize be established at the Académie des sciences for "the best work dealing with hypnotism, suggestion, and, generally, with physiological actions that could be exercised at a distance on the animal organism."[43] Although the idea was well received by the Académie des sciences, the establishment of such a prize also required the acceptance of the Ministère de l'instruction publique. In February 1905, the Conseil d'état announced its decision to deny the creation of the Jacques de Reinach

Foundation by the Académie des sciences. The refusal was justified by the claim that the planned foundation "did not appear to have a purely scientific purpose."[44] The affair could have come to a stop there, but in 1910 a new request was made. By then, Madame de Reinach had died, but her daughter still dreamed of establishing a scientific prize. In a letter to the Académie des sciences, she expressed her disappointment at what had happened and renewed her request: "Because of the kind welcome that your respectable group had given to the project of the creation of a prize in the name Jacques de Reinach . . . I allow myself to renew this offer that we had been forced to give up, the Conseil d'état having thought better than to accept it."[45] Juliette de Reinach reformulated her wish, but modified the original offer: the donation would be increased to 50,000 francs, and the name of the foundation would be changed to "Fanny Emden," her mother's maiden name. This modification, she explained, would prevent the controversy attached to the name of Reinach. In 1892, Jacques de Reinach, a banker and a financier, had been implicated in the Panama scandal and discovered dead in his bed following a warrant for his arrest (suicide had been suspected). With the change to a less publicized name, the request was accepted. By July 1910, the ministry had authorized the Académie des sciences to establish the Fanny Emden Foundation.[46]

If a prize was now established, members of the academy certainly seemed reluctant to grant it. In January 1910, the prize was discussed at a session of the Académie des sciences. Three thousand francs would be given every two years.[47] In 1911, the commission in charge of giving the prize refused to award it, explaining its motives in a report: "It seems to us that the prize should, after all, be given only to a work that has made new facts known and, especially, given absolutely undeniable proofs of these facts."[48] Having failed to find such work, the commission did think that two books in particular, although they had not escape the general reproach, did deserve recognition: Émile Boirac's *La psychologie inconnue* and Julien Ochorowicz's *La suggestion mentale*. It recommended that 2,000 francs be given to Boirac and 1,000 francs to Ochorowicz as encouragement.[49] In 1913, the commission opted for the same approach, recommending *Sur quelques réactions au contact de la plaque photographique* by Guillaume de Fontenay for 2,000 francs and Jules Courtier's *Rapport sur les séances d'Eusapia Palladino à l'Institut général psychologique* for 1,000 francs, but once again, the prize was not awarded.[50] In 1918, the money was given to Albert Dastre's widow in memory of her husband, a member of the Académie des sciences and writer of *La vie et la mort*, a decision that was approved by Juliette de Reinach herself.[51]

The prize was finally given for the first time in 1919 to Léon Chevreuil for his book *On ne meurt pas*. That year, Émile Boirac was mentioned again, this time for *L'avenir des sciences psychiques*.[52] In 1923, the prize was awarded for a second time to Georges-René-Marie Marage for *Ce qu'il faut penser des sourciers*.[53] In 1929, the prize was given to César de Vesme for his *Histoire du spiritualisme expéri-mental*.[54] That was the last time the prize was mentioned in the *Compte rendus* of the Académie des sciences (except for a reference in a list of physiological prizes in 1939.)[55] News of the Fanny Emden prize had initially ignited optimism within in the psychical research community, but the early excitement had quickly dissipated once it became clear that it did not amount to much. Except for the prize given to de Vesme in 1929, the honor was never given to any major figure in the psychical research community. As such, it is likely that the community rapidly understood that this prize would not bring about the kind of interactions they so wished with the larger scientific community. It was only mentioned on occasion in the journals of psychical research and never seems to have been given much credence.

During the 1920s, as metapsychics and the IMI entered the scene, a potential collaboration between members of the institute and the world of the universities became more likely. In the spring of 1922, four psychologists from the Sorbonne decided to undertake a series of experiments to determine the authenticity of ectoplasm. "[W]e could not refuse scientific examination to phenomena that, as strange as they appear in our present state of knowledge, are considered real by serious observers," they later wrote.[56] From March to June 1922, the psychologists met with the medium Eva C. in the physiological laboratory of the Sorbonne for fifteen séances. They hoped to witness the production of ectoplasm. Even more than the 1905–7 experiments with Eusapia at the IGP, the 1922 experiments with Eva at the Sorbonne constituted an effort to take metapsychists and their work seriously.[57]

It was the journalist Paul Heuzé who had first conceived of the Sorbonne ex-periments. He wanted to organize a series of séances witnessed by well-respected psychologists with no ties to the metapsychical community. By the 1920s, Heuzé had built a reputation as a crusader against the excesses of psychical researchers and their mediums. He took it upon himself to convince a few scientists who were respected in their field and did not believe in ectoplasm to meet an accomplished medium for a few séances. If they could be convinced of the authenticity of ecto-plasm, he reasoned, then the world would have to accept their reality. Heuzé re-solved to ask Henri Piéron, at the time professor of physiological psychology and director of the laboratory of physiological psychology at the Sorbonne, to organize a series of séances with one of the most convincing mediums of the time. Piéron was an ideal choice: although he did not believe in the reality of mediumistic phenomena, he was willing to consider them with an open mind. He was also a respected psychologist, renowned for his attempts to build an objective psychology and rid it of introspection. With three other members of the faculty, Georges Dumas, a professor of experimental psychology and pathology, Louis Lapicque, a professor of physiology, and Henri Laugier from the physiology laboratory, Piéron accepted Heuzé's invitation and prepared to hold the séances in his laboratory.[58]

While Heuzé had no trouble finding respected psychologists to participate, he had more difficulty persuading his chosen medium to perform for them. Eva Carrière — or Eva C. as she was often referred to — was none other than Marthe Béraud, of the Villa Carmen in Algiers, famous for the materialization of the spirit Bien-Boâ. In 1909, under the pseudonym Eva Carrière, she had embarked on a successful second career as a medium. Denying earlier accusations of fraud and distancing herself from her past spirit materializations, she had acquired a reputation for her ectoplasms and had begun a long-lasting association with the respected psychical researcher Juliette Bisson.[59] In 1912, the two women had even decided to live together in order to facilitate mediumistic observations. Over the next ten years, and in large part thanks to Bisson's work, Eva succeeded in retaining her popularity in the psychical movement as the most famous French medium. When first approached by Heuzé, however, both she and Bisson showed extreme reluctance to participate in the séances. For years, psychical researchers had been asking for serious consideration by institutional science, but when finally faced with the possibility, Bisson was apprehensive, fearing failure and ridicule, or success and the appropriation of her work by the psychologists. Having been reassured by Heuzé on the intentions and openness of the four scientists, however, she agreed to participate. Eva and Bisson would come to the physiological laboratory of the Sorbonne, where a series of séances would be performed to determine the authenticity of her ectoplasms.

Before the séances could begin, Piéron's laboratory had to be modified according to Bisson's instructions. Mediumistic phenomena were sensitive and numerous conditions were required for their successful production. Unfortunately, those were the same conditions that would render observation and experimentation very difficult and facilitate fraud. Nonetheless, Piéron obliged and gave instructions to build a dark room, with a small section separated by black curtains for the medium. He installed a rheostat to dim the lighting, noting that all of this had reduced visibility by half for most of the room and by 90 percent inside the medium's small cabinet.[60]

To rule out fraud, the set of measures that would be used to control Eva was thoroughly discussed and agreed upon before the séances began. Over the years, Bisson had developed specific methods to observe her medium. At first, Eva had only accepted two photographic apparatuses and one red electric lamp in the room during a séance. By 1913, however, she had come to accept as many as eight cameras, three of them in the dark room with her. She had also agreed to have the room lighted by six red electric lamps. Over the years, the two women had become respected for their serious approach to the phenomena. In 1913, César de Vesme remarked on the good faith of the medium and her willingness to conform herself to the requirement of her experimenters: "Examination of the nose, the ears, the hair, and the throat by physicians, even a gynecological and rectal check; she withstood everything with courage."[61] During her séances with Bisson, this type of examination was standard. Eva would then be given a black leotard to wear. A lab coat would subsequently be sown over the suit to ensure she could hide nothing in her sleeves.[62]

At the Sorbonne, the searches would not be as thorough. It was agreed that, at the beginning of each séance, Eva would undress and put on a black leotard, leaving only her hands, neck, and head uncovered. Her mouth, nose, and hair would be examined. She would then be taken by the hands to the room where the séance would be held. There, Bisson would put her in a trance by taking her thumb and gazing at her for a few seconds. She would give Eva's hands to the controllers who, from then on, would not let go. The medium would be placed behind a curtain—as the phenomena were sensitive to light—but her hands and feet would remain visible at all times.[63]

By the end of March, everything was finally ready. On March 21 at 4 P.M., Eva arrived at the Sorbonne for her first séance in the modified laboratory. The meetings continued until June 23, when, at Bisson's request, they ceased. In the end, participants came to the conclusion that nothing much had happened during the fifteen séances. Most of the time, they had witnessed nothing more than the medium's agitation and violent trance state, sometimes accompanied by the production of a small amount of water near her mouth, probably saliva. On a few occasions, Eva had predicted the arrival of the phenomenon but to no avail.

The most spectacular occurrence had been the apparition of a gray substance on two occasions. On April 3, it had come out of Eva's mouth, but had been swiftly reabsorbed by the medium before any prolonged observation could be made. At quick glance, the substance had looked like a thin disc surrounded by a soft material and soaked in mucus. Whereas Bisson had seen a human face forming in the disc, Piéron and Dumas had seen nothing of the sort. Then, on May 29, after a wait

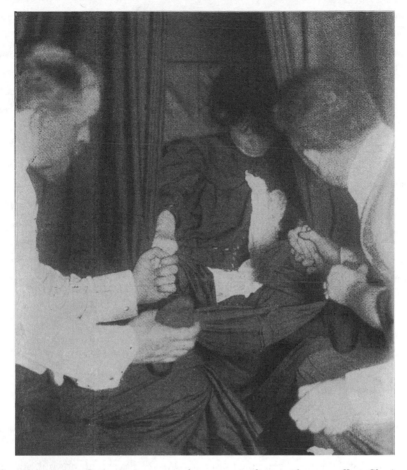

The medium Eva C. appearing to produce an ectoplasm with some effort. She is sitting behind a curtain and held by both hands and feet by experimenters on each side of her. From Juliette Alexandre-Bisson, *Les phénomènes dits de matérialisation. Étude expérimentale* (Paris: Félix Alcan, 1921), fig. 25.

of an hour and a half, Eva, feeling nothing coming, had asked to be dressed. Once in the dressing room, however, she had declared that the phenomenon was now coming. Brought back to the laboratory, she had produced a small and thin substance in her mouth, later described by Piéron and Laugier as a sheet of rubber a few millimeters long. Again, the substance had been promptly reabsorbed. If ectoplasm there was, the psychologists concluded, it had been very small, always formed in Eva's mouth, and had been extremely difficult to observe. A substance coming out of the medium's mouth after what looked like prolonged efforts to vomit, a substance without any mobility, which was rapidly swallowed back and observed with insufficient lighting could not of course be deemed sufficient evidence. For the four psychologists, nothing about the sightings warranted more than simple physiological explanations.[64]

The published report of the experiments did not surprise Bisson but she read it with frustration. Although Heuzé and the psychologists of the Sorbonne had maintained a high opinion of her integrity throughout the séances, they had cast doubt on her medium's abilities. In light of such minimal results, Bisson certainly understood their skepticism, but she expressed disappointment at the bad faith and the lack of monitoring that had been exercised by the psychologists of the Sorbonne. More tests were needed to rule out the possibility of vomit. For example, on the two occasions when she did produce a small quantity of some substance, Eva could have been forced to throw up. An inspection of the content of her stomach would have determined with certainty whether or not the phenomenon had been produced by the regurgitation of a substance ingested prior to a séance. A strong coloring fluid could also have been administered to her at the beginning of every séance to determine if the substance came from her stomach or really did materialize. X-rays could have been taken to ensure that nothing abnormal was occurring in Eva's body. Such measures, which Eva would have willingly undergone, would have contributed to the identification of the substance. Instead, the psychologists had remained obsessed with explanations involving forced vomit, with little intention of changing their minds. For the lack of results during the séances, Bisson blamed the four psychologists. They had come in late, showed impatience, and made too much noise. Eva had been in good health and should have been able to provide impressive ectoplasms but, according to Bisson, no phenomena could have been produced in such conditions.[65]

The most important problem with the Sorbonne séances, however, was the nature of the ectoplasmic phenomenon itself. When faced with the possibility of a fair investigation by respected scientists, Bisson had been reticent, not because she did not believe in the reality of ectoplasm—every account of her character emphasizes her integrity—but, most likely, because she realized the problematic nature of the phenomena she was studying. While commenting on her experiments at the Sorbonne, she explained: "The phenomena that I study do not enter into what is called 'scientific explanation,' because they cannot be reproduced at will and remain in the domain of observations only . . . they have to be caught when they come."[66] These were not simple phenomena to observe, nor were they easy to come by. This was still a burgeoning field, with a weak theoretical framework, which could account for any outcome at a séance. If nothing occurred, it was because of the conditions. Even fraud could be explained by the medium's strong wish to produce a phenomenon at any cost. This kind of attitude was not convincing for Piéron and his colleagues, but this was how psychical researchers and metapsychists had built their own field. If Bisson had shown reluctance to participate in the Sorbonne séances, it was probably because she realized that her chosen approach was not compatible with that of the psychologists and would eventually lead to a disappointing outcome, no matter what Eva produced.

Bisson was not the only psychical researcher who expressed discontent at the report of the experiments. Geley, then director of the IMI, wrote that the Sorbonne had not discredited ectoplasm. Eva's abilities had already convinced numerous observers; her ectoplasms had been seen, photographed, even touched on previous occasions. Her failure in front of the psychologists could easily be explained by the unsympathetic atmosphere of the laboratory.[67] For Geley, the report's main interest lay in the fact that in their dealings with ectoplasm, the professors of the Sorbonne had used the same methods of observation and control as those of metapsychists. Although they had failed to perform a thorough investigation, the psychologists had accepted the methods of research of previous observers, which implied the validation of the results that had already been obtained by the IMI and the larger metapsychical community.[68]

News of the report, however, led to an assault on metapsychics and psychical research in the French press. The experiments were mocked in journalistic accounts, in which the small gray substance became chewing gum that Eva was swallowing and regurgitating at will. It seemed to metapsychists that journalists took every opportunity offered to ridicule their field.[69] The IMI felt compelled to strike back and organized a series of new séances with the medium Jean Guzik. The published report of the séances was signed by thirty-four observers. It confirmed their belief in the mediumistic ability to move objects at a distance and create the sensation of touch without contact.[70] Following the "Rapport des trente-quatre," Heuzé summoned a group of five scientists, including Laugier (who had participated in the Sorbonne experiments) and the renowned physicist Paul Langevin, to observe Guzik for themselves. This time, something did happen: objects moved without apparent cause and sensations were felt, but observers remarked that the phenomena, which always appeared in close proximity of the medium, had occurred when they had been either distracted or tired. From this, they concluded that the medium somehow had managed to free one of his legs from the control of his neighbors to produce the different phenomena. When new séances were held under tighter control, nothing happened.[71] The opportunity for metapsychics' to gain greater acceptance seemed to be vanishing rapidly.

The Sorbonne experiments and their follow-ups suggest that the relationship between institutional and marginal science are complicated and not always antagonistic. Piéron, Dumas, Lapicque, and Laugier may not have acted according to Bisson's wishes, but they were willing to consider mediumistic phenomena and continued to keep an open mind even after the Sorbonne séances were done. Piéron invited both Eva and Guzik to come to his laboratory for further séances, but neither of them took him up on his invitation. Although no more séances took place in the Sorbonne laboratory, it was not because of the refusal of the psychologists of the institution to receive mediums. Psychical researchers wished for the approval of scientists. What they did not express was their need for this acceptance to come on their own terms. Bisson and others did not want psychologists to appropriate the phenomena; they wanted their own way of approaching séances to be taken seriously. This wish to be accepted with their own methods and their own set of explanations, however, often prevented them from being taken seriously by others. The case of the Sorbonne experiments illustrates this tension well.

METAPSYCHICS AT THE INTERNATIONAL LEVEL

The IMI had been created as an international institute that would provide metapsychists with a means to interact and collaborate across societies and nations. It was intend both to bring greater visibility and credibility to the discipline and also to improve its prospects for scientific acceptance. By the time of the IMI's creation in 1919, there had already been international interactions and collaboration in the psychical research community. In 1905, for example, Charles Richet had accepted the presidency of the Society for Psychical Research, becoming the second non-Briton to hold the position after William James.[72] In 1913, the presidency was again offered to a Frenchman, Henri Bergson. The philosopher expressed both surprise and pleasure when accepting the title. He was, by his own admission, not an ideal candidate. If he had an interest and a belief in psychical phenomena, it was not because he had studied them, but because their existence was compatible with his philosophical beliefs. In the acceptance speech he gave in London, Bergson emphasized the possibilities psychical research presented for the hypothesis of the survival of the soul. With time, scientists would become accustomed to the idea that consciousness was not limited to the physical organism. Proof would follow; new laws governing the spiritual would be discovered. A vitalistic biology would arise to chart the invisible internal forces that governed the sensible forms of living organisms.[73] The choice of Bergson for president, although clearly motivated by his intellectual respectability outside psychical research circles, might also have been favored because of the international character it gave the SPR. In 1925, the SPR awarded its presidency to a third Frenchman, Camille Flammarion. Again, the decision was likely motivated in part by Flammarion's international reputation. Thus, by the 1920s, efforts had been made to promote international relations in psychical research, but the initiative had come mostly from the British, who, until then had dominated the field.

From August 26 to September 2, 1921, the First International Congress of Psychical Research was held in Copenhagen, with researchers coming from Great Britain, Germany, the United States, Belgium, Denmark, Sweden, Finland, Holland, Czechoslovakia, Russia, Latvia, Peru, and France to present and share their findings. From France, Geley and Sudre represented the IMI, with Juliette Bisson, Émile Magnin, G. Mélusson, Gabriel Delanne, and Guy du Bourg de Bozas also attending. Geley opened the congress with a letter from Richet (who was unable to attend), in which the famous Nobelist urged the audience to focus on facts rather than theory, on science rather than religion. The general secretary of the congress, Carl Vett, a Danish businessman, welcomed participants to Copenhagen, emphasizing the importance of the French participation in the international enterprise: "It has always been the great virtue of the French nation to clear new paths to human knowledge; it seems that in this domain as well France will keep its place at the front."[74] The participants returned home after having passed a resolution to encourage scientific research on psychical phenomena that would illuminate some of the fundamental problems of psychology, declaring "that the aim of psychical science must be to eliminate everything that is not authentic and prepare the way for the incorporation of well-established phenomena into scientific knowledge."[75]

Participants at the 1921 Copenhagen congress agreed on the need for a second international congress of psychical research to take place in two years' time. IMI delegates suggested Paris, but Warsaw was chosen. From August 29 to September 5, 1923, the Second International Congress of Psychical Research was thus held in the Polish capital. Participants came from Germany, Austria, Denmark, Spain, the United States, Great Britain, Holland, Iceland, Italy, Latvia, Norway, Sweden, Czechoslovakia, Turkey, Poland, and France. Once again, the IMI sent Geley and Sudre. Bisson discussed the Sorbonne experiments of the previous year; and Richet, absent again, was unanimously voted honorary president.

Where enthusiasm had characterized the Copenhagen proceedings, the Warsaw congress was marked by caution and a greater need to separate psychical research from religion or doctrines. On the first day of the congress, a resolution emphasizing the scientific character of the event was unanimously adopted:

> The second International Congress of Psychical Research
> — Protests against the confusion of spiritism with psychical science that occurs daily in all countries;
> — Declares that the hypothesis of human survival is only one possible interpretation of the facts, and that in the actual state of knowledge, no interpretation can be regarded as proven;
> — Affirms once again the positive and experimental character of psychical science outside of all moral and religious doctrine.[76]

In addition, the Polish committee asked that psychical research be defined more clearly, that a demarcation between natural and supernatural phenomena be made, and that supernatural phenomena be distinguished as mediumistic, metapsychic, or parapsychic and classified as such. A commission made up of Pierre Lebiedzinski of Poland, Albert von Schrenck-Notzing of Germany, and René Sudre of France was formed to consider these propositions. On the last day of the congress, the commission presented its conclusions. Inter alia, it declared:

> The congress considers that this term [psychical research], introduced by the English SPR, is too vast for the set of phenomena that it studies, but it has no objections to retaining it as long as it is understood to apply to the phenomena known in France as *métapsychiques* and in Germany as *parapsychologiques* and *parapsychophysiques*. . . . More generally, it is difficult for the Congress to impose a valid terminology for all countries; it can only recommend that scholars not increase the difficulties of research by unnecessarily creating new terms when the old ones already suffice.[77]

Not all suggestions were adopted, however, and tensions began to surface between those participants who were concerned with delimiting the territory of psychical research and those who felt that the field should remain open to all. Carl Vett, general secretary and instigator of the Copenhagen congress, for example, deplored the fact that although anyone could attend the congress, only those invited by their national committees could read papers and discuss them. Vett felt that this practice provided an unrepresentative sample of the psychical research done in each of the participating countries.[78]

In Warsaw, a new congress was planned for 1926 in Florence. It was in Paris, however, that a congress was held in 1927. Just as in the two previous instances, participants had to be invited by national committees or by the organizing committee itself to be able to attend the proceedings in Paris. Papers read at the congress had to be preapproved by a national committee. Finally, reports had to refer to experimental research and be "inspired by *pure science*."[79] In his report on the preparations for the Third International Congress of Psychical Research, Osty strongly emphasized that one of the best ways to attract the attention of the "scientific elite" was to conduct the congress entirely and unconditionally in a scientific manner.[80]

For the members of the IMI, who had been growing increasingly unsatisfied by the persistent presence of spiritism within their discipline, a scientific congress would be possible only if the strictest selection of participants was made. Attendees at the congress should be "exclusively people of science used to presenting and accepting only what is provable, and to reach the explanation of phenomena by experiment only."[81] At the Paris congress, the centralizing tendencies of the IMI became more apparent than ever. Osty was pushing for the development of an international vocabulary of metapsychics: "We need a vocabulary marking the end of a diversity of words to designate the same things, of a multiplicity of interpretations for certain words, of the confusion resulting from terms having another meaning in the ordinary language from which they have been taken."[82] On its own territory, the IMI pushed for greater cohesion under its organization. By 1927, it was now apparent that the IMI was not successful in the mission it had given itself to provide a centralizing force to the field of metapsychics or international psychical research. At the 1927 congress, tensions and frustrations were clearly mounting.[83]

The 1921 congress in Copenhagen had been the first occasion for psychical researchers and metapsychists to meet in such great numbers. At the time, the Danish committee had proposed that a permanent organization be created in Copenhagen to prepare future international congresses. Geley and Sudre, who had been sent as the official representatives of the IMI, both protested against this project claiming that the international community of psychical researchers already possessed such an organization in Paris, the IMI. They asked participants at the congress not to divide their efforts into two international organizations. In the end, however, it was decided that the congress would assign to one representative from each participating nation the task of forming a national committee of three members, who would then relay all requests and suggestions on future congresses to the Danish committee, serving as the link between all committees.[84] The members of the IMI felt this responsibility should have been given to them. This was after all the role they had assigned themselves.

Between 1921 and 1927, Santoliquido and a few others worked toward a new strategy. At the Paris congress, Santoliquido made two requests: first, that the decision to hold the next congress in Athens be postponed for a while and, second, that he be included in the new Comité supérieur des Congrès, alongside Richet, Lodge, and Driesch. As both of his propositions failed to get unanimous support in the audience, Santoliquido withdrew his demands. When asked by Osty why he had wished to be on this committee, he answered: "This . . . would have been useful to me for the goals that I set myself; I will make them known to you when they will become tangible."[85] A few months later, Osty recalled, Santoliquido had revealed to him:

> You will now understand the meaning of my requests to the Congress. I had hoped that the participants knew me enough to know that it was not petty, personal motives that dictated my requests. But another disappointment will not make much difference. In Geneva, seat of fashionable society, I have just prepared an International Permanent Center of meetings where metapsychical researchers will be able to meet each other for conferences and congresses in such a serious atmosphere in which no scholar will find anything to blame. Things will happen in it as in the better-organized scientific congresses.[86]

Santoliquido asked Osty to join him in this project.[87] In April 1928, Osty wrote to Driesch about the plan for a permanent center of the congresses in Geneva: "Attached, I send you the presentation of a project that is currently ripe for rapid execution. It consists, as you will see, in accomplishing the last necessary act so that our congresses become truly scientific meetings in their preparation and their execution."[88] This enterprise was clearly a measure taken against the ways in which Vett was organizing the international congresses, leaving participation as open as possible: "Such is not the opinion of the scientific world that observes us and would rapidly line up at our side when we have shown it that we are people of method and precise vision without the pollution of extra-scientific elements."[89] A Permanent Congress Center in Geneva would bring about the rapid interaction of the metapsychical community with the scientific world, Osty argued.[90]

At this point, only Richet, Santoliquido, Osty, and a few professors in Geneva knew about the project. In a letter to Richet written on the same day, Osty explained his reasons for promoting this new organization: "The only individual who remains capable of a possible initiative is Carl Vett, a charming and very pleasant man, friend of department store directors and theosophists, and in no way familiar with the need for metapsychics to cross over into accepted science."[91] For Osty, it had become clearer with each passing international congress that if metapsychics were ever to become an accepted science, it would have to be presented in a different manner:

The aim of all scientific personalities who take an active interest in metapsychics is to introduce this branch of science into the universally accepted and taught science as soon as possible. For this, our successive congresses can be of great use.

But it is evident—and the preparation and subjects of the last congress have made it clear to me—that competent, full-time direction is missing from the preparation, the execution, and the application of the decisions of our congresses.[92]

By the late spring of 1928, the project was taking a more concrete turn. The first meeting of the center took place on June 13 in Geneva. A provisional committee was made up of about a dozen members, including renowned psychical researchers Driesch, Lodge, Richet, Osty, and Santoliquido, as well as Giovanni Ciraolo, president of the Italian Red Cross, and a few scientists from Geneva, such as Charles Baudouin, director of the Institut de psychologie et de psychothérapie; the psychologist Édouard Claparède (co-editor of the journal *Archives de psychologie* with his cousin Théodore Flournoy); the educator Adolphe Ferrière, founder of the progressive education movement and the International League for New Education; and the anthropologist Eugene Pittard.[93] Problems and confusions over the creation of the center quickly arose. In early July, Driesch wrote Osty about the concerns of many on the purpose of the new center. Driesch had received letters from Schrenck-Notzing, Vett, and a few English psychical researchers deploring his participation in the Geneva project. Driesch now felt that the situation had not been made clear to him before he had agreed to participate in the center. In fact, he was beginning to understand that the Geneva center had been created as an independent foundation with no attachment to the present series of international congresses, a fact he deplored as creating a competing institution and promoting a schism in European metapsychics.[94]

Twelve days later, Osty replied quite clearly:

We cannot be perpetually condemned to contend with congresses in which anyone can come and say anything, which the lack of scientific direction dooms to inevitable mediocrity, under the pretext of competition, schism, or other myths. If we want metapsychics to appear to the scientific world as a science at last, let us have the courage to create something that is rigorous scientific work.

The Geneva creation has the ambition to be a very serious organization attracting only genuine competencies. It does not replicate what exists. It does not wish in any way that the present congresses be eliminated.

Personally, before knowing that a center of conferences and congresses of psychical research would be established in Geneva, I had informed [Carl] Vett, after the Paris congress, of my resignation as a member of the congresses, having been able to judge the incredible defectiveness of this organization, which was, in the end, only a disparate meeting more damaging than useful to the reputation of metapsychics.

When I was made aware of the project to create a permanent center in Geneva, I agreed to collaborate right away, hoping that in it, at last, there would be a possibility to do serious work in an environment fitted for this. And it was only after having obtained the certitude that it was something that was very serious and feasible that I wrote to you to ask you to sit on the initial organizational committee.[95]

In a letter to Schrenck-Notzing written that same day, Osty showed surprise at his colleague's reaction: "It is not possible that you would not understand the interest that there is in the existence of meetings for genuine metapsychical researchers, meetings firmly excluding all *the amateurism* that stick to this science and compromises it."[96] Schrenck-Notzing quickly replied that he would have no objections to the center if understood as an extension of the international congresses and including the members of the national committees: "You talk, in your letter, of genuine metapsychical researchers that must be reunited in Geneva; but the names that you have so far cited to me, with the exception of French researchers, are completely unknown in our field."[97] Two days later, Schrenck-Notzing pursued his critique of the Geneva center, accusing its participants of attempting to impose French hegemony on the international metapsychical community: "All this appears to be only a helping hand to ensure the international direction of metapsychics to the francophone nations. Do you really take us for so blind that we could not discern the true personal motives for this preference?"[98] Osty received letters from Lodge and Richet, both dated July 23 and announcing the refusal of each man to participate in an enterprise surrounded by such controversy and intrigue. Richet further added that he could never be on the same side as René Sudre and his schemes.[99]

By August 1928, a new dispute had arisen. The announcement had been made in the *Revue métapsychique* that an international congress would be held in Geneva in October 1930; unfortunately, the Athens congress was set to take place just before that. Schrenck-Notzing wrote Osty accusing the Geneva center of dictatorship and sabotage.[100] Answering Schrenck-Notzing's claim that the Paris congress had been the one in which the official participation had been at its weakest, Osty alleged that the members of the IGP had refused to participate for political reasons.

Having been invited to the congress, Courtier had replied that he and the members of the IGP did not wish to attend an international congress of psychical research because they did not want to have contacts with Germans. As for academics, their absence could be explained, in large part, by the fact that the congress had been held during university recess, a lucky coincidence for Osty: "Personally, I have been very satisfied by this, for the taste of academics for metapsychics would certainly not have been enhanced by the nonsense uttered by some of the participants at the congress."[101] Between Osty and Schrecnk-Notzing, a lot more than the correct way to practice metapsychics was at stake. The ectoplasm research with which Schrenck-Notzing had been associated was a source of embarrassment for Osty. The latter even wrote to Richet: "Schrenck's only possible motive is my resistance to his desire that the frauds of Marthe Béraud [Eva C.] caught on photos be [regarded as] genuine materializations. He has had recourse to *less than delicate* means to force me to hold for practically true what was false in the past."[102]

The purpose of the Geneva center became even more evident when the possibility of canceling the Athens congress in order to avoid dividing the community was brought up. For Osty, there could be no obligations toward the national committees of psychical research. Participation in the Geneva congress would be based on personal invitations only: "I remind you, the Committee of the Geneva Center wishes to preserve complete freedom in its individual convocations. . . . That is to say that it refuses to recognize the National Committees."[103] A congress in Geneva was never meant to replace or even complement a congress in Athens. It was to function independently of other organizations.[104] The congress in Geneva was set for October 1930. By June of that year, however, it was moved back to March 1931. By November 1930, Osty, who had just learned that Santoliquido was dying, began to question the future of the Geneva congress.[105] By December 18, 1930, after exchanging a few letters back and forth about the possibility of holding the congress, Osty received a letter from Geneva informing him of the lack of enthusiasm for the Swiss center and the decision to abandon the project initiated by Santoliquido.[106]

As for the Athens congress, it was held as planned in 1930, with Driesch as honorary president. No French or Italian representatives attended. Without members of the IMI there, representatives of the SPR offered to publish the proceedings of the conference in English (until then, all proceedings had been published in French) and to hold a congress in London in 1932. English concerns regarding the survival of the soul dominated the Athens congress, something of which the French would not have approved. Upon opening the meeting, Vett declared: "Without being suspected of spiritism, we have almost all of us, I think, become convinced through psychical research of the survival of the soul after death."[107] Without IMI representatives, the tone had certainly changed. At the congress, news of the creation of a permanent center in Geneva was clearly a cause for concern. After considering a possible collaboration with the Swiss center, it was decided, however, that the nucleus of another permanent international organization would be established in London, this one open to all who wished to participate. The IMI's hopes of domination seemed to be all but quashed.

In the end, the idea of a permanent center of metapsychics in Geneva free of any damaging influence was perhaps an impossible dream for a would-be science still in the process of defining itself. As Schrenck-Notzing wrote to Osty:

> In the explanations of your letters, you do not take into consideration the fact that parapsychology is not yet, in any way, a genuine science, but that it is merely trying to become one. We cannot completely exclude the profane (amateur) element, as is the case in congresses of ethnology and anthropology. We are still, in the case of mediums, witnesses, and observers of occult phenomena, reduced to the collaboration and the testimony of individuals who do not hold academic titles. Your institute itself has followed this principle.[108]

Perhaps it was too soon then to organize such a center. In any case, the failed attempt to create a permanent center in Geneva marked the end of the IMI's ambition to dominate metapsychics at the international level.

∽∾

While preparing for the 1927 International Congress of Psychical Research, Osty had asked himself why progress was so slow in coming.[109] In fifty years, metapsychics (and psychical research before it) had accomplished very little when compared to other sciences. In fact, it still remained largely unacknowledged in the scientific world. For those who studied it, it had continued to be a discipline under construction, a possible future. Although it lent itself to collaboration with psychology, physiology, and the physical sciences, it stood isolated in its approach to metapsychical phenomena. By 1927, members of the IMI like Osty could not help but be disheartened in the face of so little progress at incorporating their field into the scientific corpus. They were annoyed that, after all these years, metapsychics was still crippled by an association with spiritism and occultism, that it was still the subject of ridicule. After all, they were creating a highly complex and important science, a task that should deserve the respect and consideration of their peers.

It is difficult to characterize the emergence and legacy of the IMI in simple terms. It was an effort to imitate science in order to become scientific, but while continuously glorifying a marginal status. As much as metapsychists were working toward the acceptance of metapsychical phenomena by the sciences, as much as they were pushing for their accounts to become the definitive explanation on the phenomena, they were also relishing in a sense of heroism, of noble sacrifice for a lost cause. They did not want to lose control over the explanation, but neither did they want to lose their status as rebels and victims. Paradoxically, by working toward the incorporation of metapsychics into what they labeled "official science" while at the same time continuing to assert their differences from it, they were ultimately contributing to their own marginalization and ultimate demise.

Conclusion

The 1930s opened with dark clouds hovering over the Institut métapsychique international. With the deaths of Rocco Santoliquido and Jean Meyer, the institute lost two of its most dynamic and valuable leaders. By then, Gabriel Delanne, Camille Flammarion, Gustave Geley, and most of the initial advocates of psychical research and metapsychics in France were gone. The few that were left would die within a decade: Charles Richet in 1935 and Eugène Osty in 1938. The IMI never really recovered from the disappointing results of the Sorbonne séances, or its failure to establish a permanent center in Geneva. Consequently, members of the institute abandoned their dream of creating an international metapsychics organization and playing a leading role in getting the phenomena accepted by the mainstream scientific community. The 1930s also saw the end of financial stability for the IMI. The confidence and the assurance exhibited at the institute throughout the 1920s had vanished. From a larger perspective, it was also the last time an attempt was made in France to define psychical and other supernatural phenomena as worthy of scientific attention.

Interest in supernatural phenomena had by no means died down. Séances, mediums, and fortunetellers continued to occupy a niche in popular culture, but the lofty scientific undertones that had marked the preceding decades were rapidly disappearing. Visiting Paris in the 1920s, the Hungarian journalist Cornelius Tabori described a city in which interest in occult and unexplained phenomena permeated daily life. He wrote of streets filled with "occult mishmash," slot machines offering printed palm readings, a horde of "miracle-doctors, visionaries, fortune-tellers, palmists, radio-mechanical interpreters of dreams, phrenologists, hypnotists, clairvoyants, telepaths, dowsers, hypersensitives, psychological trainers, mind-readers, magnetic-electric soul-searchers, learned astrologers, graphologists and magicians" proposing their services in the privacy of the customer's home, or through the post for those living in the countryside.[1]

The city described by Tabori was remote from the seriousness of Kardec's séances or the IMI's laboratory and measuring instruments. Outside the walls of the IMI, the world of supernatural manifestations was evolving even as metapsychists were beginning to organize themselves. By the 1930s and 1940s, the trend had become more pronounced. Spiritists and occultists did not cease to exist, but they turned inward and no longer promoted their interpretations of séances and other phenomena outside their own circles. Today, séances continue to be organized in Paris, and spiritism is still practiced. The tomb of Allan Kardec at the Père-Lachaise cemetery is visited frequently, and vases filled with flowers can always be found there. As for psychical researchers and metapsychists, they did not abandon their work. The IMI (which still exists today) continued to publish the *Revue métapsychique* until 1982.[2] Until 1938, it pursued its initial activities with increasing financial difficulties under the direction of Eugène Osty. The 1940s brought a new generation of enthusiasts to the institute. It marked a revival of metapsychics in France, but this resurgence was accompanied by a greater withdrawal from the scientific scene. Metapsychists became more critical of the academic world of the universities and academies for its inflexibility and its refusal to consider unconventional approaches and theories. The trend was now toward a more mystical metapsychics and an increased interest in the mysteries of the Orient. Alongside the *Revue métapsychique*, other journals dedicated to the unexplained appeared in France during this period.[3] Through all of this, however, the IMI was not without its difficulties, and, in 1955, financial troubles forced it to scale back its activities and move to more modest location in Paris.[4]

During the 1930s, France played little part in international psychical research. The term *métapsychique*, which had never gained popularity outside the country, now seemed forgotten. In 1955, the first international congress of parapsychology was held in Holland. It marked the beginning of a redefinition of the subject of research. Gone were ectoplasm and the attempt to prove the survival of the soul. Parapsychologists moved away from both the more spiritual and spectacular claims of the past. According to them, the phenomena were productions of the mind, not of the soul, and the methods of research were to be those of experimental psychology. In western Europe and North America, parapsychology found a crack in the door of the universities. At Duke University in North Carolina in the 1930s, a group of researchers including Joseph Banks Rhine studied extrasensory perception

(ESP), perceptions believed to have been received outside the range of the known senses. In an experimental setting and using statistical analysis, Rhine tested volunteer students using cards and dice. Through his articles and books, he did much to publicize parapsychology, and he continued his work at Duke until his retirement in the 1960s. By then, much had changed. In 1969, the Parapsychological Association, originally proposed by Rhine and founded in the 1950s, became part of the American Association for the Advancement of Science, leading to a greater willingness to consider the topic in academic settings. In 1979, the Princeton Engineering Anomalies Research Lab was launched at Princeton University. At the University of Edinburgh, the Koestler Parapsychology Unit was created in 1985. Today, there are still a number of privately funded laboratories around the world in which ESP and similar phenomena are studied, including the IMI, now enjoying a resurgence under the direction of an experimental psychologist specializing in parapsychology after decades of decline. If it remains at the margins of science, parapsychology has nonetheless succeeded in ways that both psychical research and metapsychics have not.[5]

ஐ-௦

Investigating the supernatural is not a simple, straightforward task. It involves phenomena that are elusive and challenging to observe and puts researchers in danger of being ridiculed by others for their perceived eccentric interests. In 1954, a potential scandal involving the early members of the IMI illustrated these kinds of difficulties. In an article first published by the Society for Psychical Research in Britain, the metapsychist Rudolf Lambert revealed details of an encounter that had led him to discover a fraud covered up by many of the early members of the IMI years before. In 1927, following the Paris congress, he had learned that the French committee had excluded Juliette Bisson from its proceedings because of the discovery of an old note from Geley implying her fraudulent participation in some of Eva C.'s materializations.[6] Confronting Osty (then director of the institute) about the claim, Lambert had learned of the whole affair. He revealed:

> When I saw Osty at the Institut the next morning, he showed me many stereoscopic pictures in Geley's files. One could clearly distinguish that the various materializations were artificially attached to Eva's hair, in parts by the hair itself, which must have represented at times the hair of the materialization, in parts with wires (textile or metal), which Eva's partisans would undoubtedly have claimed were also materialized. . . . Osty also told me that he had wanted to publish this discovery. As, however, Richet and Schrenck Notzing had opposed

this energetically, and M. Jean Meyer, the militant spiritist who financed the *Institut métapsychique*, had also demanded that the scandal be kept secret, Osty had to abandon the idea to make his discovery public. But after exercising a certain degree of pressure, he had succeeded in excluding Eva's procurer, Mrs. Bisson, from the French Committee of the Congress.[7]

Lambert had waited more than twenty years to reveal his secret, which he claimed put Geley and of course Juliette Bisson's work in a suspicious light. Since Eva Carrière had been searched on numerous occasions without any artifacts being found on her, the evidence seemed to suggest the fraudulent complicity of Bisson.[8]

In 1955, extracts of Lambert's confession were published in the *Revue métapsychique*, followed by a defense of Bisson's integrity written by René Warcollier, then president of the IMI. Warcollier emphasized Bisson's own doubts in her book *Les phénomènes de matérialisation* and the fact that Eva's productions had been rare. If Bisson had in fact been the mediums' accomplice, Warcollier argued, it would have been easy for the two women to produce more successful séances.[9] Except for this short reply, Lambert's revelations fell on deaf ears in France until 1968, when René Pérot, a friend of Bisson's in her old age, published his recollections in the *Revue métapsychique*. Bisson had left the world of metapsychics long before Lambert's accusations had been made public. She had mostly ignored the episode, claiming to be resigned to the lack of comprehension of her work, Pérot wrote. Until her death in 1956, she had remained convinced of the authenticity of the phenomena she had witnessed, and, more than ten years after the death of his old friend, Pérot still felt the need to defend her honor and integrity.[10]

This episode suggests how hard it was to remain impartial when observing such uncanny phenomena, even for an honest researcher. There is no denying that fraud occurred at séances, and that it occurred frequently. The evidence against Eusapia and Eva, among others, is convincing, but a case against the good faith of Bisson, Geley, Richet, and others is more difficult to make. They can be accused of blind enthusiasm and naïveté, of stubbornness and lack of rigor, but it is hard to see then as insincere and deceitful. Feeling impelled by their mission, and convinced of the authenticity of the phenomena, could they at times have seen what they wanted to see?[11] Is it possible that in their eagerness to establish the authenticity of the mediums, they themselves became unsuspecting actors in a performance? Psychical researchers based their confidence in the phenomena in part on the conviction that, although it was difficult to remain impartial when confronted with the spectacular manifestations produced by mediums, it was possible to control for fraud before séances began. In their writings, they always gave great credence to the goodwill of mediums like Eusapia and Eva who consented to be thoroughly searched, observed, and controlled. But, one could argue that such searching and probing was also to the advantage of the mediums, particularly with researchers like Bisson, Geley, and Richet who had a vested interest in the phenomena. If successful, a collaboration with respected researchers could enhance a medium's credibility in the eyes of his or her larger audience, paying or otherwise.

Beyond the spectacular nature of the phenomena, other difficulties were always present. The unpredictable and capricious nature of the supernatural meant that it could rarely be investigated at the more typical sites of scientific experimentation or be easily duplicated. Even for psychical researchers and metapsychists who sometimes tried to approach the manifestations in more conventional settings, the limitations imposed by the phenomena were tremendous. For Sudre, any effort to legitimize metapsychics by attempting to reproduce the phenomena in the austere environment of the laboratory was pointless. In the end, séances were and would always remain séances:

> In the very commendable wish to get rid of the mystical or socialite character of the research, we have copied the laboratories of physics or physiology, we have designed a big, bare room, with the floor and walls completely tiled, into which we have crammed iron or copper instruments of bizarre shapes, photographic chambers, electric apparatus, etc. When the subject enters such a place, he has the impression of entering a surgical clinic or a torture room, and this is enough to sterilize him.[12]

Like most people in his field, Sudre believed that a successful séance required a favorable atmosphere: "A study room, a room furnished with taste and inhabited is often the best metapsychic laboratory."[13] Séances could not be reproduced at will in test tubes and beakers, in bleak laboratories, hospital rooms, or any of the normal settings of scientific research. When Sudre wrote these lines in 1926, this indisputable fact had been a fundamental problem plaguing every attempt to study this and other extraordinary phenomena for decades. As mainstream science had developed an internal division of labor, and the methods and approaches of its internal subdisciplines gained increasing acceptance, groups of reported phenomena had been left behind. By nature, the investigation of the supernatural could be constraining and frustrating. Phenomena were unreliable and never easy to produce or reproduce; they were fleeting and more likely to belong to the private sphere, to the world of women and the calm and comfort of the home, rather than that of men and the laboratory. If for some, all of this put them outside of the scope of scientific investigation, for others, it simply constituted a serious challenge to scientific exploration and research, one that could lead to a radical change in our understanding of human potential, our spiritual selves and the material world. Such an enterprise, they thought, was worth the struggle.

Between the 1850s and the 1930s, scientists attended séances, considered occult teachings, worried about the mental health of spiritists, and tested mediums. Beyond the hospitals and university laboratories, science enthusiasts and autodidacts actively contributed to a widespread exploration of the supernatural as believers, followers, witnesses, participants, and researchers. Their stories, interactions, and sometimes even collaborations present a portrait of French science that reaches beyond the walls of respected and recognized institutions and extends away from secular and materialistic approaches to include the many men and women who partook in discussions on the possibilities and limits of science exploration. They reveal a rich and vibrant diversity of unorthodox beliefs and practices that existed in the broader scientific culture of the period. They show that French science was not strictly defined and controlled by its universities and learned academies, that there was always space for challenges, oppositions, and creative investigations. The investigation of the supernatural in France permeated the broader reaches of the scientific culture of the period, ranging from conversations with spirits around tables, where spiritists hoped to communicate with the dead and obtain proof of the survival of the soul, to occultists seeking a system of knowledge based on ancient revelations, to physicians, psychiatrists, and psychologists dismissing alleged supernatural experiences as pathological and psychical researchers and metapsychists attempting to create a new field of research and a new science of unknown forces in humans.

Although spiritists, occultists, psychical researchers, and metapsychists deplored the exclusiveness of science, their experiences indicate that alternative visions were present in the sciences and, at times, even considered within the most select of institutions. If anything, they show us that rejection came more from the fact that the phenomena with which they were preoccupied were by nature difficult to examine using the conventional methods of scientific exploration than from a failure on the part of scientific institutions to allow spaces for alternative knowledge. Each of these groups was, to an extent, given some consideration by scientists. Their ideas were examined. The call of spiritists and occultists for fundamental revisions of contemporary science was never going to get them very far. Psychical researchers and metapsychists, however, adopted a more reasonable approach and were marginally successful. The fact that some of them continued to hold positions at universities even while promoting the existence of psychical phenomena suggests the complicated associations of both fields with the more established scientific institutions. If, in the end, physicians, psychiatrists, and psychologists were more successful at imposing their explanations of the supernatural, this story nonetheless suggests that scientific marginality was not rejection, but that it was never acceptance either; it remained somewhere in between, never in but not completely out of the scientific landscape of the time.

Acknowledgments

I am very grateful for the support I received from numerous corners for this project. For their financial and institutional support, I am indebted to the Social Sciences and Humanities Research Council, the University of Notre Dame, the Max Planck Institute for the History of Science, and the University of Guelph. For the generous use of their archives, I want to thank Jacques Pernet and Patrick Fuentes at the Observatoire Flammarion, and Mario Varvoglis at the Institut métapsychique international.

Numerous colleagues and friends have read and commented on some version of this work over the years. Many thanks to Tara Abraham, Joanne Astley, Doris Bergen, Charlotte Bigg, Uljana Feest, Frédéric Gosselin, Christopher Hamlin, David Allen Harvey, Elizabeth Hayes, Jenna Healey, Don Howard, Christopher Laursen, Linda Mahood, and John Monroe for their help through what at times has seemed like a never-ending process. I have greatly benefited from the patience and the support of my two dissertation advisors, Michael Crowe and Thomas Kselman. I am also grateful to Bob Brugger and everyone at the Johns Hopkins University Press for their guidance through the publication process, and to two anonymous readers for their thoughtful and very useful comments.

Over the years, many friends have encouraged me in one way or another. Among them, I extend my warmest thanks to David Aubin, Alisa and Peter Bokulich, Linda Cambell, Martine Caron, Catherine Carstairs, Brandon Fogel, Darin Hayton, Sébastien Marineau, Andreas Mayer, Stuart McCook, Annika Parrance, Stefan Petri, Karen Racine, Steve Ruskin, Florian Sennlaub, Stephan Williams and everyone else who has supported me along the way. A very special thank you to Segire, Louis Morel l'Horset, and Louis Steven Morel l'Horset for continuing to make me feel like part of your family year after year.

Finally, I am immensely grateful for all the love and support of my family. My parents, Réjean Lachapelle and Margo Nobert, my brother, Ugo Lachapelle, my partner, Francis Poulin, and my daughter, Ariane Lachapelle-Poulin, are a constant source of happiness and inspiration in my life.

⌒⌒

Chapter 5 of the book appeared in an earlier version as "Attempting Science: The Creation and Early Development of the Institut métapsychique international in Paris, 1919–1931," *Journal of the History of the Behavioral Sciences* 41 (2005): 1–24.

Notes

INTRODUCTION

1. Figuier wrote numerous books on science. See, e.g., Louis Figuier, *Les grandes inventions modernes dans les sciences, l'industrie et les arts* (Paris: Hachette, 1912); *Les races humaines* (Paris: Hachette, 1872), *Les merveilles de la science ou description populaire des inventions modernes*, 4 vols. (Paris: Hachette, 1867–91), and *Les mystères de la science*, 2 vols. (Paris: Hachette, 1892–93). On science and life after death, see Louis Figuier, *Le lendemain de la mort, ou la vie future selon la science* (1871; new ed., Paris: Hachette, 1912).On the ambiguous boundaries between science and magic in popular works of science in the second half of the nineteenth century, see Sofie Lachapelle, "Science on Stage: Amusing Physics and Scientific Wonder at the Nineteenth-Century French Theatre," *History of Science* 47 (2009): 297–315.

2. On the rich, mystical character of the second half of the nineteenth century in France, see Ruth Harris, *Lourdes: Body and Spirit in the Secular Age* (London: Allen Lane, Penguin Press, 1999) and "Possession on the Borders: The 'Mal de Morzine' in Nineteenth-Century France," *Journal of Modern History* 69 (1997): 451–78; and Thomas Kselman, *Death and the Afterlife in Modern France* (Princeton, NJ: Princeton University Press, 1993) and *Miracles and Prophecies in Nineteenth-Century France* (New Brunswick, NJ: Rutgers University Press, 1983). On a case of stigmata in French-speaking Belgium at the time, see Sofie Lachapelle, "Between Miracle and Sickness: Louise Lateau and the Experience of Stigmata and Ecstasy," *Configurations* 12 (2004): 77–105.

3. Jean-Martin Charcot, "La foi qui guérit," *Revue hebdomadaire* (1892): 112–32. On Janet's work, see, e.g., Pierre Janet, *De l'angoisse à l'extase: Études sur les croyances et les sentiments*, 2 vols. (Paris: Alcan, 1926–28), "Les états de consolation et les extases," *Journal de psychologie* (May 1925): 369–420, "La psychologie de la croyance et le mysticisme," *Revue de métaphysique et de morale* 43 (1936): 327–58, 507–32; 44 (1937): 369–410, and "Un cas de possession et l'exorcisme moderne," *Bulletin de l'Université de Lyon* 2 (1895): 41–57. On the role of physicians at the Lourdes sanctuary, see Jason Szabo, "Seeing Is Believing? The Form and Substance of French Medical Debates over Lourdes," *Bulletin of the History of Medicine* 76 (2002): 199–230. On the role of mystics in the development of late nineteenth-century psychiatry, especially at La Salpêtrière, see Jean Céart, "Démonologie et démonopathies au temps de Charcot,"

Histoire des sciences médicales 28, 4 (1994): 337–43, Jacqueline Lalouette, "Charcot au coeur des problèmes religieux de son temps," *Revue neurologique* 150, 8–9 (1994): 511–16, Jacques Maître, "De Bourneville à nos jours: Interprétations psychiatriques de la mystique," *Évolution psychiatrique* 64 (1999): 765–78, and *Une inconnue célèbre. La Madeleine Lebouc de Janet* (Paris: Anthropos, 1993).

4. Maurice Crosland, *Science Under Control: The French Academy of Sciences, 1795–1914* (New York: Cambridge University Press, 1992)

5. On the popularization of science in France, see *La science pour tous: Sur la vulgarisation scientifique en France de 1850 à 1914,* ed. Bruno Béguet (Paris: Bibliothèque du CNAM, 1990), and *La science populaire dans la presse et l'édition, XIXe et XXe siècles,* ed. Bernadette Bensaude-Vincent and Anne Rasmussen (Paris: CNRS, 1997).

CHAPTER 1: FROM TURNING TABLES TO SPIRITISM

1. *La Patrie,* May 27, 1853, cited in *Examen raisonné des prodiges récents d'Europe et d'Amérique, notamment des tables tournantes et répondantes par un philosophe* (Paris: Librairie de Vermot, 1853), pp. 11–12.

2. There are numerous contemporary accounts, with a few variations, of the "Rochester rappings," as they were often called. For a contemporary French perspective, see Louis Figuier, *Histoire du merveilleux dans les temps modernes,* vol. 4: *Les prodiges de Cagliostro, les magnétiseurs mystiques, l'électro-biologie, la fille électrique, les escargots sympathiques, les esprits frappeurs, les tables tournantes et les médiums, les spirites,* 3rd ed. (Paris: Hachette, 1881), pp. 273–76; and Jacques Babinet, "Les sciences occultes au XIXe siècle, les tables tournantes et les manifestations prétendus surnaturelles considérées au point de vue de la science de l'observation," *Revue des deux mondes* 6 (1854): 510.

3. For a history of the American movement, see, e.g., Ann D. Braude, *Radical Spirits: Spiritualism and Women's Rights in Nineteenth-Century America* (Boston: Beacon Press, 1989), and Laurence R. Moore, *In Search of White Crows: Spiritualism, Parapsychology, and American Culture* (New York: Oxford University Press, 1977).

4. Reprint of an article from *L'univers,* July 26, 1852, in *Examen raisonné des prodiges récents d'Europe et d'Amérique, notamment des tables tournantes et répondantes par un philosophe* (Paris: Librairie de Vermot, 1853), p. 28.

5. According to Janet Oppenheim, Mrs. Hayden and Mrs. Roberts, two American mediums, arrived in 1852 and 1853 respectively while the famous D. D. Home arrived from America in 1855. Janet Oppenheim, *The Other World: Spiritualism and Psychical Research in England, 1850–1914* (Cambridge: Cambridge University Press, 1985), p. 11.

6. Guillard, *Table qui danse et table qui répond. Expériences à la portée de tout le monde* (Paris: Garnier frères, 1853), p. 7.

7. Figuier, *Histoire du merveilleux,* 4: 312.

8. Ferdinand Silas, *Instruction explicative des tables tournantes d'après les publications allemandes, américaines, et les extraits des journaux allemands, français et américains* (Paris: Houssiaux & Dentu, 1853), pp. 27–28.

9. D., "Les tables volantes," *L'illustration*, June 25, 1853.

10. Figuier, *Histoire du merveilleux*, 4: 337

11. *Chez Victor Hugo. Les tables tournantes de Jersey. Procès-verbaux des séances*, ed. Gustave Simon (Paris: Louis Conard, 1923); and John Warne Monroe, *Laboratories of Faith: Mesmerism, Spiritism, and Occultism in Modern France* (Ithaca, NY: Cornell University Press, 2008), pp. 50–52.

12. Henri Carion, *Lettres sur l'évocation des esprits à Madame *** (Paris: Dentu, 1853), p. 7.

13. Adolphe Leboucher, *Nouvel abracadabra, ou le choléra et les tables tournantes* (Paris: Simon Raçon, 1854), e.g., describes a séance in which the spirits recommended warding off cholera with sage.

14. On French mediums and particularly women and mediumship, see Nicole Edelman, *Voyantes, guérisseuses et visionnaires en France, 1785–1914* (Paris: Albin Michel, 1995).

15. Although the commission never submitted an official report, Jacques Babinet, who was a member, wrote on the affair explaining Cottin's abilities as a simple muscle trick. See Babinet, "Les sciences occultes au XIXe siècle," pp. 512, 515–17. On Cottin and other teenage girls, see Edelman, *Voyantes, guérisseuses et visionnaires*, pp. 76–77; Figuier, *Histoire du merveilleux*, 4: 160–87; and Dr. Tanchou, *Enquête sur l'authenticité des phénomènes électriques d'Angélique Cottin* (Paris: Germer Baillère, 1846).

16. *Guide des plaisirs à Paris, Paris le jour, Paris la nuit. La tournée des grands-ducs. Comment on s'amuse. Ou l'on s'amuse. Ce qu'il faut voir. Ce qu'il faut faire* (Paris, 1906), p. 27.

17. Nicole Edelman and Bertrand Méheurst both emphasize this association of spiritism with animal magnetism in France by considering spiritism as a later development of animal magnetism: Edelman, *Voyantes, guérisseuses et visionnaires*, and Méheust, *Somnambulisme et médiumnité*, 2 vols. (Le Plessis–Robinson, Paris: Institut Synthélabo pour le progrès de la connaissance, 1999).

18. Jacquet, *Les tables tournantes* (Nantes: W. Busseuil, 1853).

19. Article in the *Revue contemporaine*, May 1853, reproduced in *Examen raisonné*, p. 58.

20. Silas, *Instruction explicativedes tables tournantes*, p. 28.

21. P.-F. Mathieu, *Un mot sur les tables parlantes suivi du crayon magique et du guéridon poète* (Paris: Jules Laisné, 1854), pp. 23–24.

22. Article in the *Gazette médical de Strasbourg*, May 1853, reproduced in *Examen raisonné*, p. 59.

23. *Avis sur les tables tournantes et parlantes par un écclésiastique* (Paris: Devarenne & Perisse frères, 1853), pp. 10, 18–19.

24. Jules-Eudes de Mirville, *Pneumatologie. Des esprits et de leurs manifestations fluidiques. Mémoire adressé à l'Académie par J.-E. de Mirville*, 4th ed. (Paris: H. Vrayet de Surcy, 1858), p. 58. For accounts of the Ciderville hauntings, see Babinet, "Les sciences occultes au XIXe siècle," p. 513, and Figuier, *Histoire du merveilleux*, 4: 239–43. The occultist Papus indirectly discusses the events of Ciderville in a pamphlet dealing with a case of haunting in Valence-en-Brie. See Gérard Encausse [Papus], *La*

maison hantée de Valence-en-Brie, étude critique et historique du phénomène (Paris: Édition de "l'Initiation," 1896), pp. 1–31.

25. André Pezzani, *Condamnation des manifestations spirites* (Paris: Ledoyen librairie, 1860), p. 6.

26. Ibid., p. 9.

27. Carion, *Lettres sur l'évocation*, p. 86.

28. P.-F. Mathieu, *Un mot sur les tables parlantes*, p. 24.

29. Abbé Almignana, *Du somnambulisme, des tables tournantes et des médiums, considérés leurs rapports avec la théologie et la physique* (Paris: Dentu & Germer-Baillière, 1854).

30. De La Giroudière, *Observations sur les tables tournantes* (Dijon: Loireau-Feuchot, 1853), p. 2.

31. Ernest Bersot, *Mesmer, le magnétisme animal, les tables tournantes et les esprits*, 4th ed. (Paris: Hachette, 1879), pp. 75–76.

32. "Recherches expérimentales sur les tables tournantes, par M. M. Faraday," *L'illustration*, July 9, 1853.

33. Michel-Éugène Chevreul, "Lettre à M. Ampère," *Revue des deux mondes*, 2nd ser., 2 (1833): 258–66.

34. Michel-Éugène Chevreul, *De la baguette divinatoire, du pendule explorateur et des tables tourantes, au point de vue de l'histoire, de la critique et de la méthode expérimentale* (Paris: Mallet-Bachelier, 1854).

35. Jacques Babinet, "Sciences des tables tournantes au point de vue de la mécanique et de la physiologie," *Revue des deux mondes* 5 (1854): 410.

36. Darius-C. Rossi, *Les tables tournantes et parlantes* (Toulon: Veuve Baume, 1853); Armand Maizière, *Sur les tables tournantes* (Reims: E. Luton, 1853). Other mechanical accounts include D. Agar de Bus, *Lettre au sujet des tables tournantes* (Issoudun: H. Cotard, 1853), and J. Assézat and H. Debuire, *Magnétisme et crédulité, ou Solution naturelle du problème des tables tournantes* (Paris: Garnier frères, 1853).

37. Babinet, "Les sciences occultes au XIXe siècle," p. 530; *Grosjean à son évêque, au sujet des tables tournantes* (Paris: Ledoyen, 1854), pp. 4–8, *Seconde lettre de Grosjean à son évêque au sujet des tables parlantes, des possessions, des sibylles, du magnétisme et autres diableries* (Paris: Ledoyen, 1855), p. 55; Pierre Janet, *L'automatisme psychologique*, 4th ed. (Paris: Alcan, 1903), pp. 377–401.

38. On Chevillard's position, see Monroe, *Laboratories of Faith*, p. 175.

39. A. Chevillard, *Études expérimentales sur le fluide nerveux et solution rationelle du problème spirite* (Paris: E. Dentu, 1882), p. 19.

40. Louis-Alphonse Cahagnet, *Arcanes de la vie future dévoilés*, 3 vols. (Paris: Germer Baillière, 1854).

41. L. de Guldenstubbé, *Pneumatologie positive et expérimentale. La réalité des esprits et le phénomène merveilleux de l'écriture directe démontrés* (Paris: A. Frank, 1857).

42. For example, *Plan proposé pour l'amélioration de l'instruction publique* in 1828, *Cours pratique et théorique d'arithmétique* in 1831, *Manuel des examens pour les brevets de capacité* in 1846, and *Cathéchisme grammatical de la langue française* in 1848.

43. Henri Sausse, *Biographie d'Allan Kardec* (1896; Paris: Pygmalion, 1993), pp. 20–27.

44. Ibid., pp. 33–34.

45. Kardec named his doctrine "spiritisme" to differentiate it from "spiritualisme," a more general philosophy in France at the time postulating that mind and body are separate. In France, "spiritualisme" contributed to the development of the science of psychology. On this see, e.g., John I. Brooks III, "Philosophy and Psychology at the Sorbonne, 1885–1913," *Journal of the History of the Behavioral Sciences* 32 (1996): 379–407.

46. Kardec, *Esprits*, pp. 86–87.

47. Ibid., pp. xvii, 66–67, 93.

48. Allan Kardec, *Le Livre des médiums*, 49th ed. (Paris: Librairie des sciences psychiques et spirites, n.d.), pp. 463–65; Sausse, *Biographie d'Allan Kardec*, p. 43.

49. Lynn L. Sharp, "Rational Religion, Irrational Science: Men, Women, and Belief in French Spiritism, 1853–1914" (Ph.D. diss., University of California, Irvine, 1996), p. 19.

50. In a police report on a banquet held in Kardec's honor in Toulouse in February 1862, only fourteen attendants are mentioned. Archives nationales, F19 10927. On his travels to promote spiritism, see Allan Kardec, *Voyage spirite en 1862* (Paris: Vermet, 1988; http://pagespersorange.fr/charles.kempf/Livres/vs.htm [accessed August 20, 2010]).

51. Allan Kardec, *Le livre des esprits* (Paris: Philman, 1999), p. xxxix, and id., *Le livre des médiums*, pp. 458–67.

52. Kardec, *Le livre des esprits*, pp. xviii–xix.

53. Ibid., pp. 14–25, 74, 292.

54. Kardec, *Le livre des médiums*, pp. 21, 30–31.

55. Kardec, *Le livre des esprits* pp. 18–19 and 68–69.

56. For a discussion of the association of ideas on extraterrestrial worlds with the spiritist doctrine, see Michael Crowe, *The Extraterrestrial Life Debate, 1750–1900* (Cambridge: Cambridge University Press, 1986).

57. Camille Flammarion, *Mémoires biographiques et philosophiques d'un astronome* (Paris: E. Flammarion, 1911), pp. 180–81.

58. Ibid., p. 99.

59. Ibid., p. 187.

60. Camille Flammarion, *Les maisons hantées; en marge de «La mort et son mystère»* (Paris: E. Flammarion, 1923), p.35.

61. Camille Flammarion, letter to Charles Burdy, June 18, 1861, Fonds Camille Flammarion de l'Observatoire de Juvisy-sur-Orge, MS copybook, "Miscellanées 1861."

62. Camille Flammarion, letter to Abbé Collin, October 7, 1861, ibid.

63. Camille Flammarion, letter to Charles Burdy, October 15, 1861, ibid.

64. Camille Flammarion, letter to Allan Kardec, November 12, 1861, ibid.

65. Camille Flammarion, *Les habitants de l'autre monde; révélations d'outre-tombe publiées par Camille Flammarion, communications dictées par coups frappés et par l'écriture médiumnique au salon Mont-Thabor, médium mademoiselle Huet* (Paris: Ledoyen, 1862), p. 7.

66. "Les frères Davenport," *Revue spirite* 8 (1865): 311–21.

67. Camille Flammarion, *Des forces naturelles inconnues; à propos des phénomènes produits par les frères Davenport et par les médiums en général, étude critique par Hermès* (Paris: Didier, 1865); and id., *Mémoires biographiques*, pp. 319–21.

68. Flammarion, *Mémoires biographiques*, pp. 495–98. On Flammarion, see Philippe de La Cotardière and Patrick Fuentes, *Camille Flammarion* (Paris: Flammarion, 1994).

69. Flammarion, *Mémoires biographiques*, p. 498.

70. On Kardec's succession, see Lynn L. Sharp, *Secular Spirituality: Reincarnation and Spiritism in Nineteenth-Century France* (Lanham, MD: Lexington Books, 2006), pp. 73–83.

71. "But, objet et raison d'être de cette institution," *Bulletin mensuel de la société scientifique d'études psychologiques* 2 (1883): 63.

72. Ibid., p. 64.

73. *Revue spirite*, 1858, 1888.

74. Dr. T. Puel, *Revue de psychologie expérimentale* 1 (1875): 194–95.

75. Henri F. Ellenberger, *The Discovery of the Unconscious: The History and Evolution of Dynamic Psychiatry* (New York: Basic Books, 1970), pp. 338–42.

76. Gabriel Delanne, "Revue de l'année," *Revue scientifique et morale du spiritisme*, January 1898: 387.

77. Gabriel Delanne, *Le phénomène spirite, témoignage de savants*, 6th ed. (Paris: Leymarie, 1909), i.

78. Ibid., p. 264.

79. Ibid., p. ii.

80. Ibid., p. 295.

81. Gabriel Delanne, *Le spiritisme devant la science* (1885; Paris: Bibliothèque de philosophie spiritualiste, 1923), p. 241.

82. Ibid., p. 117.

83. Ibid., pp. 321–22.

84. Gabriel Delanne, *La réincarnation* (1924; Paris: Vermet, 1985).

85. Émile Boirac, *L'étude scientifique du spiritisme* (Paris: Durville, 1911), p. 6.

86. Ibid., pp. 20–21.

87. J. Malosse, *La vie après la mort. La science de l'âme, preuves scientifiques de la survivance, les vies successives, les mondes et l'univers* . . . (Lyon: Oeuvre populaire philosophique, 1924), p. 4.

88. L. Chevreuil, *On ne meurt pas. Preuves scientifiques de la survie* (Paris: Jouve, 1935), p. vii. Others held similar opinions. See A. Bernard, *Le monde invisible. Les manifestations des esprits: Actions psychiques. Apparitions. Matérialisations. Tables tournantes. Écritures et dessins automatiques. Le spiritisme devant la science. Les faits expérimentaux* (Paris: Henri Durville, 1921); Dr L.-Th. Chazarain, *Les preuves scientifiques de la survivance de l'âme* (Paris: Henri Richard, 1905); D. Isnard, *La réalité des phénomènes spirites. Le spiritisme considéré au point de vue scientifique et philosophique* (Paris: N. Trécult, 1920); and Thomas Mainage, *Immortalité, entretiens sur le*

problème de la survivance; l'univers, les religions, l'homme, la métapsychique, la conscience, la raison, éternité, 2nd ed. (Paris: P. Téqui, 1926).

89. Chevreuil, *On ne meurt pas,* pp. 7–8.

90. Pierre-Camille Revel, *Le hasard, sa loi et ses conséquences dans les sciences et en philosophie, suivi d'un essai sur la métempsychose d'espèce basée sur les principes de la biologie et du magnétisme physiologique. Compendium* (Paris: Durville, 1926), pp. 78–79.

91. Gustave Geley, *Reincarnation,* trans. Ethel Archer (London: Rider, [1929?]), pp. 30–31.

CHAPTER 2: OCCULT WISDOMS, ASTRAL BODIES, AND HUMAN FLUIDS

1. For a general history of the French occultist movement, see David Allen Harvey, *Beyond Enlightenment: Occultism and Politics in Modern France* (DeKalb: Northern Illinois University Press, 2005); and Christopher McIntosh, *Eliphas Lévi and the French Occult Revival* (London: Rider, 1972).

2. On Lévi, see McIntosh, *Eliphas Lévi.*

3. See, e.g., Alexandra David-Néel, *Mystiques et magiciens du Tibet* (Paris: Robert Laffont, 1977); and Alexandra David-Néel, *Voyages d'une parisienne à Lhassa* (Geneva: Gonthier, 1964).

4. See, e.g., Daniel Arnauld, *Fakirs et jongleurs* (Paris: Firmin-Didot, 1889); Ernest Bosc, *Yoghisme et fakirisme hindous (Introduction à la yoga). Yoghis et fakirs: leur entraînement, leurs facultés psychiques, leurs pouvoirs* (Paris: G.-A. Mann, 1913); Louis Jacolliot, *Voyage au pays des fakirs charmeurs* (Paris: E. Dentu, 1883); Louis Noir, *Le Fakir* (Paris: Fayard frères, 1899); Louis Rousselet, *Le charmeur de serpents* (Paris: Hachette, 1910); and Sédir, *Le fakirisme hindou et les yogas,* 2nd ed. (Paris: Librairie générale des sciences occultes, Bibliothèque Charcornac, 1911).

5. Paul Gibier, *Le spiritisme, fakirisme occidental* (Paris: Durville, 1887), pp. 80–81. On the Western interest in the East, see, e.g., Mark Bevir, "The West Turns Eastward," *Journal of the American Academy of Religion* 62 (1994): 747–67.

6. The organization split after Blavatsky's death. In 1895, Olcott remained in Adyar and continued to lead the organization, while William Q. Judge (vice president of the TS and general secretary of the American section) returned to America to start his own branch of the society. Currently, both branches still exist. The first is known as "Theosophical Society–Adyar", and its headquarters are still located in Adyar. The branch that split off was originally located in New York, but its headquarters are now located in Pasadena, California. It is known as just the "Theosophical Society," though the location of the California headquarters is usually added to avoid confusion. See the official web site of the Theosophical Society (Adyar), www.ts-adyar.org, and the home page of the Theosophical Society (Pasadena, www.theosociety.org (both accessed August 20, 2010).

7. For a discussion of Blavatsky and the history of the Theosophical Society, see Sylvia L. Cranston, *The Extraordinary Life and Influence of Helena Blavatsky, Founder*

of the Modern Theosophical Movement (New York: Putnam, 1993)and Alex Owen, *Place of Enchantment: British Occultism and the Culture of the Modern* (Chicago: University of Chicago Press, 2004) pp. 29–41.

8. Thomas Mainage, *Les principes de la théosophie, étude critique*, 9th ed. (Paris: Éditions de la "Revue des jeunes," 1922), p. 13.

9. Ibid., p. 12.

10. Lady Caithness, "Notre programme," *L'aurore*, December 1886: 1.

11. Comtesse d'Adhémar, "Avant-propos," *Revue théosophique*, 1889: 1.

12. Joris-Karl Huysmans, *À rebours* (1884; Paris: Garnier-Flammarion, 1993); id., *Là-bas* (1891; Paris: Garnier-Flammarion, 1993). On Stanislas de Guaïta, see Ch. Barlet, "L'oeuvre philosophique de Guaïta," *L'initiation*, January 1898: 1–21; Dr. Marc Haven, "Le kabbaliste," ibid.: 32–37; François Jollivet-Castelot, "L'alchimiste," ibid.: 58–63; E. Michelet, "L'artiste," ibid.: 48–57; Gérard Encausse [Papus], "L'oeuvre de réalisation," ibid.: 21–31; Sédir, "L'oeuvre de Stanislas de Guaïta au point de vue occulte," ibid.: 37–48.

13. On this episode, see Dr Bataille, *Le diable au XIXème siècle ou les mystères du spiritisme* (Paris: Delhomme & Briguet, 1892–95); Georges Bois, *Causeries du dimanche. Première partie. Franc-Maçonnerie, occultisme, polémique avec «Le Diable au XIXe Siècle»* (Paris: Victor Retaux, 1897); Diana Vaughan, *Mémoires d'une expalladiste parfaite initiée, indépendante* (Paris: Librairie antimaçonnque, publication mensuelle, 1895–97); id.,, *La restauration du paganisme. Transition décrétée par le sanctum regnum pour préparer l'établissement du culte public de Lucifer. Les hymnes liturgiques de Pike. Rituel du néo-paganisme* (Paris: Librairie antimaçonnique, A. Pierret, 1896). For a historical account, see Eugen Joseph Weber, *Satan franc-maçon: La mystification de Léo Taxil* (Paris: R. Julliard, 1964).

14. Christophe Beaufils, *Joséphin Péladan: 1858–1918: Essai sur une maladie du lyrisme* (Grenoble: J. Millon, 1993); and René-Louis Doyon, *La douleureuse aventure de Péladan* (Paris: La Connaissance, 1946).

15. Ernest Bosc, *Diabolisme et occultisme* (Nice: Chamuel, 1896), pp. 11–12.

16. Gérard Encausse [Papus], *La réincarnation. L'évolution physique, astrale et spirituelle, l'esprit avant la naissance et après la mort*, 4th ed. (1912; Paris: Dangles, 1953), p. 34.

17. Gérard Encausse [Papus], "Sociétés d'initiation en 1889," *L'initiation*, April 1889: 1.

18. Ch. Barlet, "Sciences occultes: L'initiation," *Revue des hautes-études* 1 (1886): 14.

19. Ibid., p. 15.

20. Ibid., pp. 13–19.

21. On the history of freemasonry in late-nineteenth-century France, see Mildred J. Headings, *French Freemasonry Under the Third Republic* (Baltimore: Johns Hopkins University Press, 1949).

22. Gérard Encausse [Papus], *Traité élémentaire d'occultisme. Initiation à l'étude de l'ésotérisme hermétique*, new ed. (Paris: La Diffusion scientifique, 1954), pp. 34–36.

23. *L'initiation*, May 1893: 179, reports 1,500 readers.

24. "Programme," *L'initiation*, n.d., 1888.

25. Gérard Encausse [Papus], *Enseignement méthodique de l'occultisme* (Paris: Éditions de l'initiation, 1901–2), pp. 6–10.

26. Encausse [Papus], *Réincarnation*, p. 30.

27. Ibid., p.12.

28. Ibid., p. 50.

29. Ibid., p. 92.

30. Ibid., p. 151.

31. Ibid., pp. 113–16.

32. Gérard Encausse [Papus], *La science des mages et ses applications théoriques et pratiques*, 2nd ed. (Paris: Librairie du merveilleux, Chamuel, 1892), p. x.

33. Encausse [Papus], *Traité élémentaire d'occultisme*, pp. 39–41, 124.

34. Ibid., p. 113.

35. de Meck, *Métapsychisme et occultisme. Douze conférences* (Paris: Librairie A.-M. Beaudelot, 1928), p. 5.

36. Encausse [Papus], *La science des mages*, p. x.

37. Encausse [Papus], *Traité élémentaire d'occultisme*, pp. 23–24.

38. Gérard Encausse [Papus], *La magie et l'hypnose. Recueil de faits et d'expériences justifiant et prouvant les enseignements de l'occultisme* (Paris: Éditions traditionnelles, 1975), p. 359.

39. Encausse [Papus], "Occultisme et spiritisme,"*L'initiation* (June 1890): 211.

40. Joseph Maxwell, *La magie* (Paris: Ernest Flammarion, 1922), p. 5.

41. Ibid., p. 145.

42. Ibid., p. 129.

43. Ibid., p. 272.

44. Ph. Pagnat, *L'occultisme et la conscience moderne. Enquête* (Paris: Éditions des "pages modernes," 1910), p. 30.

45. Joseph Grasset, *L'occultisme hier et aujourd'hui. Le merveilleux préscientifique*, 2nd ed. (Montpellier: Coulet et fils, 1908), p. 38.

46. *La curiosité. Journal de l'occultisme scientifique*, followed by *Revue des sciences psychiques*, 1889–98; *Écho du monde occulte. Revue bi-mensuelle de vulgarisation des sciences occultes et divinatoires*, 1905–06; *Écho du merveilleux*, 1897–1914; *L'étoile d'Orient. Revue des hautes études psychiques. Organe officiel du Centre ésotérique oriental de France*, 1908–9.

47. *Compte rendu du Congrès spirite et spiritualiste international tenu à Paris du 9 au 16 septembre 1889* (Paris: Librairie spirite, 1890).

48. Léon Denis, *Après la mort, exposé de la doctrine des esprits, solution scientifique et rationnelle des problèmes de la vie et de la mort, nature et destinée de l'être humain, les vies successives*, new ed. (Paris: Librairie des sciences psychiques, 1905), p. 10.

49. Ibid., p. 14.

50. Ibid., pp. 21–22.

51. Ibid., p. 22.

52. Ibid., p. 61–63.

53. *Compte rendu du Congrès spirite et spiritualiste international tenu à Paris du 9 au 16 septembre 1889*, p. 56.

54. Encausse [Papus], "Congrès spirite et spiritualiste international (Paris 1889)," *L'initiation*, October 1889: 2.

55. The editors, "Déclaration à nos lecteurs et à nos abonnés," *L'initiation*, February 1889): 101–2; and Le Maître, "Aux disciples," *L'étoile d'Orient* 7 (October 1908): 186.

56. Encausse [Papus], "Occultisme et spiritisme," pp. 209–12.

57. Encausse [Papus], *L'initiation*, June 1890: 206.

58. Dr. Mac Nab, "Note sur le spiritisme et les phénomènes psychologiques," *Le Lotus bleu*, 1892: 27–29.

59. Thomas Mainage (father), *Les principes de la théosophie, étude critique*, 9th ed. (Paris: Éditions de la revue des jeunes, 1922), pp. 17–18.

60. Daniel Metzger, *Spiritisme et hypnotisme* (Paris: Georges Carré, 1890), pp. 25–26.

61. Ibid., p. 31.

62. On Mesmer and Mesmerism in France, see Léon Chertok and Isabelle Stengers, *Le coeur et la raison. L'hypnose en question, de Lavoisier à Lacan* (Paris: Payot, 1989); Léon Chertok and Raymond de Saussure, *Naissance du psychanalyste. De Mesmer à Freud* (Paris: Payot, 1973); Adam Crabtree, *From Mesmer to Freud: Magnetic Sleep and the Roots of Psychological Healing* (New Haven, CT: Yale Press University, 1999); Robert Darnton, *Mesmerism and the End of Enlightenment in France* (Cambridge, MA.: Harvard University Press, 1968); Henri F. Ellenberger, *The Discovery of the Unconscious: The History and Evolution of Dynamic Psychiatry* (New York: Basic Books, 1970); and Bertrand Méheust, *Somnambulisme et médiumnité*, vol. 1, *Le défi du magnétisme* (Le Plessis–Robinson, Paris: Institut Synthélabo pour le progrès de la connaissance, 1999).

63. For example, *Annales du magnétisme animal*, 1814–16; *Archives du magnétisme animal*, 1820–23; *L'avenir médical. Journal des intérêts de tous ayant pour but la démonstration pratique du nouvel art de guérir, l'homéopathie et le magnétisme*, 1844–45; *L'hermès. Journal du magnétisme animal*, 1826–29; *Journal du magnétisme animal*, 1839–42; *Le propagateur du magnétisme animal*, 1827–28; and *Moniteur spirite et magnétique*, 1877–1900.

64. For example, A.-J.-P. Phillip (pseud. of J.-P. de Gros Durand), *Electrodynamisme vital ou les relations physiologiques de l'esprit et de la matière démontrées par des expériences entièrement nouvelles et par l'histoire reaisonnée du système nerveux* (Paris: J.-B. Baillière, 1855), posited that a magnetic force was in fact an electro-biological force.

65. Charles de Reichenbach, *Les phénomènes odiques, ou recherches physiques et physiologiques sur les dynamides du magnétisme, de l'électricité, de la chaleur, de la lumière, de la cristallisation, des actions chimiques, considérés dans leurs rapports avec la force vitale* (Paris: Librairie générale des sciences occultes, 1904); id., *Les effluves odiques, conférences faites en 1866 par le baron de Reichenbach à l'Académie I. et R. des sciences de Vienne, précédées d'une notice historique sur les effets mécaniques de l'od par Albert de Rochas* (Paris: Ernest Flammarion, 1897).

66. Armand Marc Jacques de Chastenet Puységur, *Recherches, expériences et ob-*

servations physiologiques sur l'homme dans l'état de somnambulisme naturel et dans le somnambulisme provoqué par l'acte magnétique (Paris: J.-G. Dentu, 1811).

67. For examples of fluidist accounts of the tables, see Charles Bourseul, *La vérité sur les tables tournantes* (Metz: Alcan, 1853); Joseph-Adolphe Gentil, *Magnétisme, somnambulisme. Guide du consultant et des incrédules. Tables parlantes*, 2nd ed. (Paris: E. Dentu, 1853); *Magnétisme. Moyens magnétiques pour faire tourner les tables, les chapeaux, etc. etc.* (Lyon: J. Brunet Fils, 1853); A. Noury, *L'antidote du spiritisme. Étude psychologique* (Paris: Renaud libraire, 1865); and J. Trismégiste, *Les merveilles du magnétisme et les mystères des tables tournantes et parlantes* (Paris: L. Passard, 1854). Agénor de Gasparin, *Du surnaturel*, 2 vols. (Paris: Calmann Lévy, 1892), believed the turning tables to be the product of a fluid, but the talking tables to be fraudulent. Abbé Urbain Onfray-Kermoalquion, *Lettre au sujet des tables tournantes* (St-Brieux: Prud'homme, 1855) argued that the tables reacted either to Satan (generally, talking tables) or to a human fluid (generally, turning tables).

68. Louis Goupy, *Explication des tables parlantes, des médiums, des esprits et du somnambulisme par divers systèmes de cosmologie. Suivie de la voyante de Prévost* (Paris: Germer Baillière, 1860), 201.

69. Horace Pelletier, "Expériences relatives à la force psychique,"*L'initiation*, May 1890: 124–34.

70. Paul Joire, "Étude d'une force extériorisée par l'organisme vivant et observations faites au moyen du sthénomètre," *Annales des sciences psychiques*, 1904: 243–53. Other instruments of this kind were on occasion discussed in Joire's society. See, e.g., P. Archat, "Recherche expérimentale de l'action motrice sans contact," *Annales des sciences psychiques*, June 1908: 198–201; "Le «Moteur à fluide» du comte de Tromelin," *Annales des sciences psychiques*, June 1908: 201–3.

71. Hippolyte Baraduc, *Les vibrations de la vitalité humaine. Méthode biométrique appliquée aux sensitifs et aux névrosés* (Paris: J.-B. Baillière et fils, 1904), p. v–vii.

72. Ibid., p. 3.

73. Hippolyte Baraduc, *La force curative à Lourdes et la psychologie du miracle* (Paris: Bloud, 1907), pp. iii–iv.

74. Ibid., pp. 10–12.

75. Ibid., pp. 15–18, 28.

76. "Un appareil pour photographier les esprits!" *Annales des sciences psychiques*, July 1912: 252.

77. Ibid., p. 253.

78. Gaston-Jean Mondeil, *Le fluide humain dans ses manifestations physiques révélatrices, contribution à l'étude de la radio-activité animale* (Paris: Berger-Levrault, 1921).

79. Guillaume de Fontenay, *La chimiographie et la prétendue photographie du rayonnement vital* (Paris: Société des publications scientifiques et industrielles, 1913).

80. Albert de Rochas, *L'extériorisation de la sensibilité* (Paris: Pygmalion, 1977), p. 2.

81. Albert de Rochas, *Les états profonds de l'hypnose* (Paris: Chamuel, 1892; 5th ed., Paris: Bibliothèque Chacornac, 1904), pp. 9–10.

82. Rochas, *Extériorisation*, p. 6.
83. Ibid., pp. 14–25.
84. Ibid., pp. 25–29.
85. Ibid., pp. 62–63.
86. Ibid., pp. 63–69.

CHAPTER 3: PATHOLOGIES OF THE SUPERNATURAL

1. A. Sanson reproduced in "Sur la folie spiritie. Réponse à M. Burlet, de Lyon," *RevuesSpirite. Journal d'études psychologiques* 6 (1863): 51–52.

2. Philibert Burlet, *Du spiritisme considéré comme cause d'aliénation mentale* (Lyon: Richard, 1863).

3. On the development of French psychology, see Jacqueline Carroy, Annick Ohayon, and Régine Plas, *Histoire de la psychologie en France, XIXe–Xxe siècles* (Paris: La Découverte, 2006), pp. 40–48, Laurent Mucchielli, "Aux origins de la psychologie universitaire en France (1870–1900). Enjeux intellectuels, contexte politique et stratégies d'alliance autour de la *Revue philosophique* de Théodule Ribot," *Annals of Science* 55 (1998): 263–89, and Serge Nicolas and David J. Murray, "Théodule Ribot (1839–1916), Founder of French Psychology: A Biographical Introduction," *History of Psychology* 2 (1999): 277–301.

4. A few historians have discussed the ways in which the supernatural entered both disciplines. More specifically, Pascal Le Maléfan has offered a detailed treatment of the ways in which spiritism was used in the French construction of psychopathology, Mark Micale and Ruth Harris have written on hysteria in the French Catholic context, and Régine Plas has explored the presence of the supernatural in the early development of psychology. Pascal Le Maléfan, *Les délires spirites. Histoire du discours psychopathologique sur la pratique du spiritisme, ses abords et ses avatars (1850–1950)* (Paris: L'Harmattan, 1999); Ruth Harris, "The Unconscious and Catholicism in France," *Historical Journal* 47 (2004): 331–54; Mark S. Micale, *Approaching Hysteria. Disease and Its Interpretation* (Princeton, NJ: Princeton University Press, 1994), pp. 260–84); Régine Plas, *Naissance d'une science humaine: la psychologie. les psychologues et le merveilleux psychique* (Rennes: Presses Universitaires de Rennes, 2000).

5. On Marian apparitions and the Marian revival of the nineteenth century, see Barbaro Corrado Pope, "Immaculate and Powerful: The Marian Revival in the Nineteenth Century," in *Immaculate and Powerful. The Female Sacred Image and Social Reality*, ed. Clarissa W. Atkinson, Constance H. Buchanan, and Margaret R. Miles (Boston: Beacon Press, 1985), pp. 173–200.

6. Émile Littré, "Fragment de médecine rétrospective," *La philosophie positive* 5, no. 11 (July–August 1869): 103–20.

7. On Bourneville's anticlericalism, see Bernard Brais, "Désiré Magloire Bourneville and French Anticlericalism During the Third Republic," in *Doctors, Politics and Society: Historical Essays*, ed. Roy Porter and Dorothy Porter (Atlanta: Rodopi, 1993), pp. 107–39, and Jan Goldstein, *Console and Classify: The French Psychiatric*

Profession in the Nineteenth Century (Cambridge: Cambridge University Press, 1987), pp. 362–72.

8. *Procès-verbal fait pour délivrer une fille possédée par le malin esprit à Louviers*, ed. Armand Bénet (Paris: A. Delahaye & Lecrosnier, 1883), pp. xv, xvii.

9. Jean-Marie Charcot, "La foi qui guérit," *Revue hebdomadaire*, 1892: 112–32.

10. Pierre Janet, *De l'angoisse à l'extase. Études sur les croyances et les sentiments.* 2 vols. (Paris: Alcan, 1926–28). On Madeleine and Janet, see Jacques Maitre, *Une inconnue célèbre. La Madeleine Lebouc de Janet* (Paris: Anthropos, 1993). On the political and moral motives behind the practice of retrospective medicine, see Sarah Ferber, "Charcot's Demons: Retrospective Medicine and Historical Diagnosis in the Writings of the Salpêtrière School," in *Illness and Healing Alternatives in Western Europe*, ed. Marijke Gijswijt-Hofstra, Hilary Marland, and Hans de Waardt (New York: Routledge, 1997), pp. 120–40, Jan Goldstein, "The Hysteria Diagnosis and the Politics of Anticlericalism in Late Nineteenth-Century France," *Journal of Modern History* 54 (1982): 209–39, and Patrick Vandermeersch, "The Victory of Psychiatry over Demonology: The Origin of the Nineteenth-Century Myth," *History of Psychiatry* 2 (1991): 351–63.

11. Jacques Maître, "De Bourneville à nos jours. Interprétations psychiatriques de la mystique," *Evolution psychiatrique* 64 (1999): 767–69.

12. On Morzine, see Ruth Harris, "Possession on the Borders: The 'Mal de Morzine' in Nineteenth-Century France," *Journal of Modern History* 69 (1997): 451–78. On the relationship between the Church and medicine at the end of the nineteenth century, see Hervé Guillemain, "Déments ou démons? L'Exorcisme face aux sciences psychiques (XIXe–XXe siècles)" *Revue d'histoire de l'église de France* 87 (2001): 454–62.

13. Félix de Backer, *Lourdes et les médecins* (Paris: A. Maloine, 1905).

14. Antoine Imbert-Gourbeyre, *L'hypnotisme et la stigmatization* (Paris: Bloud & Barral, 1899).

15. On Catholic physicians, see Hervé Guillemain, "Les débuts de la médecine catholique en France. La Société médicale Saint-Luc, Saint-Côme et Saint-Damien (1884–1914)," *Revue d'histoire du XIXe siècle* 26–27 (2003): 227–58.

16. Alexandre Jeanniard du Dot, *Le spiritisme dévoilé, ou les faits spirites constatés et commentés* (Paris: Bloud & Barral, 1895), p. 19.

17. Ch. Hélot, *Névroses et possessions diaboliques* (Paris: Bloud & Barral, 1897); id., *Le diable dans l'hypnotisme* (Paris: Bloud & Barral, 1899). On Hélot as a medical expert in cases of possession in Rouen, see Guillemain, "Déments ou démons?" pp. 458–59.

18. Paul Duhem, *Contribution à l'étude de la folie chez les spirites* (Paris: G. Steinheil, 1904), p. 130.

19. Marcel Viollet and A. Marie, *Le spiritisme dans ses rapports avec la folie. Essai de psychologie normale et pathologique* (Bloud, 1908), p. 40, and Dr. Paul Duhem, *Contribution à l'étude de la folie chez les spirites* (Paris: G. Steinheil, 1904) p. 99.

20. Duhem, *Contribution à l'étude de la folie*, pp. 126–30; and Pierre Boudou, *Le spiritisme et ses dangers: Quatre conférences* (Bordeaux: Féret & fils, 1921), p. 62.

21. Henry Emmanuel Marie Aubin, *Les Délires de métapsychique* (Bordeaux: Imprimerie de l'Académie et des Facultés, 1927), p. 13; Duhem, *Contribution à l'étude de la folie*, p. 18; Salo Kern, *Contribution clinique et pathogénique à l'étude des délires spirites* (Paris: Jouve, 1936), pp. 49–50; and Viollet and Marie, *Le spiritisme dans ses rapports avec la folie*, pp. 116–17.

22. Burlet, *Du spiritisme considéré*, pp. 30–57, contains a series of letters from spiritists angered by the author's claim that the practice can lead to madness.

23. On the selling of *Des Indes à la planète Mars*, see Marina Yguello's introduction in Théodore Flournoy, *Des Indes à la planète Mars. Étude sur un cas de somnambulisme avec glossolalie* (1899; Paris: Slatkine Reprints, 1983), p. 13.

24. William James and Theodore Flournoy, *The Letters of William James and Theodore Flournoy*, ed. Robert Charles Le Clair (Madison: University of Wisconsin Press, 1966), p. 90.

25. There has been some work done on Flournoy's *Des Indes à la planète Mars* `from the perspective of the history of psychiatry and psychoanalysis. See, e.g., Ellenberger, *Discovery of the Unconscious*, pp. 315–17; Olivier Flournoy, *Théodore et Léopold. De Théodore Flournoy à la psychanalyse* (Neuchâtel, Switzerland: Éditions de la Baconnière, 1986); introduction of Sonu Shamdasani in Théodore Flournoy, *From India to the Planet Mars* (1900; Princeton, NJ: Princeton University Press, 1994); preface of Hélène Boursinhac, and introduction of Marina Yaguello in Théodore Flournoy, *Des Indes à la planète Mars. Étude sur un cas de somnambulisme avec glossolalie*, 1899 ed. (Paris: Slatkine Reprints, 1983). For a firsthand account of Flournoy's life and work, see Édouard Claparède, "Théodore Flournoy. Sa vie et son oeuvre," *Archives de psychology* 18 (1923): 1–125. See also Serge Nicolas and Agnes Charvillat, "Theodore Flournoy (1854–1920) and Experimental Psychology: Historical Note," *American Journal of Psychology* 111, no. 2 (Summer 1998): 279–94. On the linguistic curiosity of new languages, see Marina Yaguello, *Lunatic Lovers of Language: Imaginary Languages and Their Inventors* (London: Athlone Press, 1991); and Mireille Cifali, "The Making of Martian: The Creation of an Imaginary Language," in Flournoy, *From India to the Planet Mars*.

26. In particular here I am thinking of William Crookes with Florence Cook and Juliette Bisson with Eva Carrière.

27. Flournoy, *Des Indes à la planète Mars*, p. ix.

28. James and Flournoy, *Letters*, p. 29.

29. Ibid., p. 47.

30. Ibid., p. 48.

31. Flournoy, *From India to the Planet Mars*, p. 14.

32. Ibid.

33. Ibid., pp. 27–30.

34. Ibid., p.14.

35. Ibid., p. 124.

36. Ibid., p. 156.

37. Ibid., pp. 135–36, 156–57; and Théodore Flournoy, "Nouvelles observations sur un cas de somnambulisme avec 'glossolalie,'" *Archives de Psychologie* 1 (1902): 101–255.

38. On the linguistic curiosity of new languages, see Yaguello, *Lunatic Lovers of Language*; Cifali, "Making of Martian."

39. Victor Henry, *Le langage martien. Étude analytique de la genèse d'une langue dans un cas de glossolalie somnambulique* (Paris: J. Maisonneuve, 1901). See also a book review in *Revue des études psychiques* (March–April 1902): 113–14.

40. Five hundred copies of the first edition were printed and a thousand each of the second and third editions, according to Flournoy, *Théodore et Léopold*, p. 166.

41. Société d'études psychiques de Genève, *Autour «des Indes à la planète Mars»* (Geneva: Georg; Paris: Librairie spirite, 1901). On the controversy that followed, see "Entre le Pr. Flournoy et les spirites de Genève," *Revue des études psychiques* (August–September–October 1901): 285–92.

42. Flournoy, *Théodore et Léopold*, p. 103.

43. Ibid., pp. 169–70.

44. Letter of Mme de Mé . . . to Flournoy, September 1, 1909. Quoted from Ibid., p. 172.

45. On Smith's new career as an artistic medium, see "Mlle Hélène Smith devient médium dessinateur," *Annales des sciences psychiques*, May 1907: 388–89.

46. Théodore Flournoy, *Esprits et médiums. Mélanges de métapsychique et de psychologie* (Geneva: Kündig; Paris: Fischbacher, 1911), pp. vi–vii.

47. Théodore Flournoy, *Le génie religieux*, 3rd ed. (Saint-Blaise, Switzerland: Foyer solidariste de librairie et d'édition, 1910), p. 13. Flournoy, preface in Émile Lombard, *De la glossalie chez les premiers chrétiens et des phénomènes similaires. Étude d'exégène et de psychologie* (Lausanne: Georges Bridel; Paris: Fischbacher, 1910).

48. Although Flournoy has remained mostly forgotten by historians, he has been somewhat rediscovered in the past twenty years by psychoanalysts who present him as a contemporary of Freud's who exerted a significant influence on Jung. See, e.g., the psychoanalyst Olivier Flournoy's analysis of his grandfather's relationship with Smith in Flournoy, *Théodore et Léopold*.

49. Pierre Janet, *L'automatisme psychologique*, 4th ed. (Paris: Alcan, 1903), pp. 401–2.

50. Dr. Eugene Azam, *Hypnotisme et double conscience. Origine de leur étude et divers travaux sur des sujets analogues* (Paris: Alcan, 1893), p. 54.

51. Pierre Janet, "Le spiritisme contemporain," *Revue philosophique* 33 (1892): 413.

52. Joseph Grasset, "Le spiritisme devant la science. À propos d'une maison hantée," *Annales des sciences psychiques*, 1903: 339.

53. Ibid., p. 41.

54. For example, Papus recounts the story of a haunting in Valence-en-Brie stopped successfully by piercing the air with a sword: Gérard Encausse [Papus], *La maison hanté de Valence-en-Brie, étude critique et historique du phénomène* (Paris: Édition de "l'Initiation," 1896), pp. 21–22.

55. Grasset, *Annales des sciences psychiques*, 1903: 85.

56. Ibid., p. 168.

57. Ibid.

58. Ibid., pp. 146–47.

59. Ibid., p. 89.

60. Ibid., p. 165.

61. Ibid., p. 103, 285.

62. Ibid., p. 178–79.

63. Ibid., p. 275.

64. Ibid., p. 268.

65. Ibid., p. 280.

66. Ibid., p. 284.

67. Allan Kardec, *Le livre des médiums*, 49th ed. (Paris: Librairie des sciences psychiques et spirites, n.d.), pp. 262–65.

68. Gustave Geley, *La psychologie anormale n'est pas explicable par la pathologie. Extrait du livre jubilaire du prof. J. Tessier* (Lyon: A. Rey, 1909).

69. Gustave Geley, *Les deux psychismes. À propos de la nouvelle théorie du professeur Grasset* (Sant Amand, Cher: Daniel-Chambon, 1903), pp. 1–10.Geley

70. Ibid., p. 13–17.

71. Pierre Janet, "Société international de l'Institut psychique," *Bulletin de l'Institut psychique* 1 (1900): 4.

72. Ibid., pp. 3–7.

73. "Réunion constitutive, 30 juin 1900," *Bulletin de l'Institut psychique international* 1 (1900): 14.

74. Janet, "Société international de l'Institut psychique," p.16.

75. *Quatrième Congrès international de Psychologie tenu à Paris du 20 au 26 août 1900. Compte rendu des séances et texte des mémoires*, ed. Pierre Janet (Paris: Alcan, 1901), p. 617.

76. Ibid., p. 656.

77. Ibid., p. 137.

78. Ibid., pp. 137–41.

79. Ibid., p. 103.

80. "Statuts de la société de psychologie," *Bulletin de l'Institut psychologique* 3 (1900): 140.

81. "Groupe d'étude de phénomènes psychiques," *Bulletin de l'Institut général psychologique* 1 (1902): 3–4.

82. Émile Duclaux, "Groupe d'études des phénomènes psychiques. Organisation d'un laboratoire," *Bulletin de l'Institut général psychologique* 4 (1902): 328.

83. César de Vesme, "L'Institut psychologique international," *Revue des études psychiques* 11 (November 1901), 321–30, recalls having paid a membership fee and hearing nothing subsequently.

84. "L'Assemblée de l'Institut général psychologique. Pas d'argent, pas de médiums," *Revue des études psychiques*, (March–April 1902): 127.

85. "Le grand développement de l'Institut général psychologique de Paris," *Annales des sciences psychiques*, March 1906: 192–93.

86. Jules Courtier, "Rapport sur les séances d'Eusapia Palladino à l'Institut général psychologique en 1905, 1906, 1907 et 1908," *Bulletin de l'Institut général psychologique* 5–6 (1908): 59.

87. Ibid., p. 415.

88. Ibid., pp. 416–19.

89. Ibid., p. 420.

90. Ibid., p. 433.

91. Ibid., p. 514.

92. Ibid., p. 519.

93. Ibid., p. 520.

94. Ibid., pp. 521–26.

95. Ibid., p. 539.

96. Ibid., p. 546.

97. Ibid.

98. Ibid., pp. 570–72.

99. Léon Demonchy, "Les rapports sur les séances d'Eusapia Palladino à l'Institut général psychologique en 1905, 1906, 1907 et 1908," *Annales des sciences psychiques* 3 (February 1909): 33–44; C. de Vesme, "À propos du rapport sur les expériences d'Eusapia à l'Institut psychologique," *Annales des sciences psychiques* 5 (March 1909): 78–85; "La discussion du rapport de M. Courtier à l'Institut psychologique," *Annales des sciences psychiques* 5 (March 1909): 92–96.

100. De Vesme, "À propos du rapport sur les expériences d'Eusapia," 79.

101. Ibid., p. 85.

102. Jules Courtier, *Post-scriptum au rapport sur les séances d'Eusapia Palladino à l'Institut général psychologique* (Vannes: Lafoyle & J. de Lamarzelle, 1929).

103. See, e.g., Louis Favre, "Dispositifs et techniques applicables aux phénomènes médiumiques d'ordre physique," *Bulletin de l'Institut général psychologique* 4 (1910): 389–412; id. "Nouveaux dispositifs pour l'étude des phénomènes médiumiques d'ordre physique," ibid. (1912): 337–43; Louis Favre, "Pourquoi faut-il étudier les phénomènes psychiques," ibid. 4 (1909): 503–26; id., "La Production des phénomènes médiumiques," ibid. (1912): 320–35; Dr Marage, "La baguette des sourciers," ibid. (1912): 317–19; B. de Rollière, "La baguette des sourciers. Classification des faits et des méthodes anciennes et modernes," ibid. (1912): 343–54.

CHAPTER 4: WITNESSING AND TESTIFYING TO PSYCHICAL PHENOMENA

1. Charles Richet, "Lettre," *Annales des sciences psychiques* 1 (1891): 4.

2. John J. Cerullo, *The Secularization of the Soul: Psychical Research in Modern Britain* (Philadelphia: Institute for the Study of Human Issues, 1982), pp. 38–45.

3. Ibid., pp. 91–94.

4. The following are listed as members of the SPR in the *Annales* in 1891: Charles Richet, Théodule Ribot, Hippolyte Bernheim, Henry Beaunis, Jules Liégeois, Ambroise Liébeault, Pierre Janet, Léon Marillier, and Xavier Dariex. Xavier Dariex, "Aperçu historique," *Annales des sciences psychiques* 1 (1891): 2.

5. Charles Richet, "Lettre," p. 7.

6. Ibid.

7. Members of the commission were Charles Richet, physician in Paris; Sully-

Prudhomme, of the Académie française; Gilbert Ballet, physician in Paris; Henri Beaunis, physician in Nancy; Colonel Albert de Rochas, administrator at the École polytechnique; and Leon Marillier, lecturer at the École des hautes-études. "Avis important," *Annales des sciences psychiques* 2 (1891): 73.

8. Félix Alcan, "Réponse à un lecteur," *Annales des sciences psychiques* 2 (1892): 64.

9. Xavier Dariex, "Les Annales," *Annales des sciences psychiques* 1 (1901): 1.

10. Ibid., p. 2.

11. *Bulletin de la Société d'études psychiques de Nancy*, 1900: 6–7; and *Bulletin de la Société d'études psychiques de Nice*.

12. For bibliographical information on César de Vesme, see René Warcollier, "César de Vesme," *Revue métapsychique* 4 (1938): 241–45.

13. Xavier Dariex, *Annales des sciences psychiques* 6 (1904): 321.

14. Warcollier, "César de Vesme," p. 242.

15. "Société universelle d'études psychiques," *Annales des sciences psychiques*, March 1910: 88.

16. Ibid.

17. Only a formal request to the effect was required. Ibid., p. 90.

18. "*Les Annales des sciences psychiques* seront dorénavant une revue illustrée bi-mensuelle," *Annales des sciences psychiques*, December 1907: 813–14.

19. The editors, "Aux Lecteurs des ANNALES," *Annales des sciences psychiques*, January 1908: 1.

20. César de Vesme, "La retraite de M. le Dr Dariex," *Annales des sciences psychiques* 7 (April 1911): 97.

21. César de Vesme, "Aux anciens abonnés des *Annales des sciences psychiques*," *Revue métapsychique*, March–April 1924: 172–73.

22. A. Bonneville, *Vision à distance. Transmission de la pensée. Les suggestions. Le fluide vital* (Algiers: Michelet, 1928), p. 4.

23. Émile Boirac, *L'avenir des sciences psychiques* (Paris: Alcan, 1917), p. 2.

24. Joseph Maxwell, *Les phénomènes psychiques* (Paris: Alcan, 1903), p. 315.

25. Xavier Dariex, "De la méthode dans les observations de la télépathie," *Annales des sciences psychiques* 1 (1891): 12.

26. Ibid., pp. 12–13.

27. Henri Durville, *La télépsychie. L'art de lire et de transmettre la pensée* (1927; Paris: H. Durville, 1973).

28. Jean Filiatre, *Hypnotisme et magnétisme, somnambulisme, suggestion et télé-pathie, influence personnelle, cours pratique* (Cosne-D'Allier: A. Filiatre, 1905).

29. Sylvain Martel, *L'art de se faire aimer de près ou de loin par l'application de la suggestion et de la télépathie, les plus puissantes influences humaines qui soient au monde*, 2nd ed., modified (Montauban: Imprimerie coopérative, 1921).

30. Paul-C. Jagot, *L'influence à distance. La transmission de pensée et la suggestion mentale. Méthode pratique de télépsychie* (Paris: Dangles, n.d.).

31. Durville, *La télépsychie*.

32. Charles Le Roy, *Le succès par la puissance mentale, cours d'application ration-nelle des forces mentales dressé à la suite d'une étude approfondie des sciences psy-*

chiques et de nombreuses expériences personnelles (Bône 1923); and Sylvain Roudès, *Vos forces mentales. Leur développement. Leur utilisation pour réussir dans la vie* (Paris: H. Dangles, 1934).

33. A. Bonneville, *Vision à distance*, pp. 21–29.

34. T.-S. Peyras, *La science occulte. Tous les phénomènes qui s'y rattachent: Hypnotisme, magnétisme, spiritisme, le pouvoir mental et la force mentale, la dualité des êtres, l'influence d'un individu sur un autre individu, la guérison des maladies par la thérapeutique suggestive. Tout le monde peut devenir hypnotiseur* (Marseille: Redon, 1906), p. 10.

35. G.-A. Mann, *La force-pensée: la faculté unique; mécanisme de la télépathie; extériorisation de la volonté; appel et captation des forces cosmiques; théorie nouvelle de l'influence de l'homme sur l'homme* (Paris: G.-A. Mann, 1910).

36. Frederic Myers, Edmund Guyers, and Frank Podmore, *Phantasms of the Living* (London: Trübner, 1886). Camille Flammarion, *L'inconnu et les problèmes psychiques; manifestations de mourants. Apparitions. Télépathie. Communications psychiques. Suggestion mentale. Vue à distance. Le monde des rêves. La divination de l'avenir* (Paris: E. Flammarion, 1900), pp. 93–94. The answers of the survey are still available today at the Fonds Camille Flammarion de l'Observatoire de Juvisy-sur-Orge.

37. Ibid., pp. 287, 290 and 581.

38. Charles Richet, *Bulletin des armées*, January 10, 1917. The results of the investigation were never published, but some of the testimonies can be found in Charles Richet, *Traité de métapsychique* (Paris: Alcan, 1922). Perhaps not surprisingly, the *Annales des sciences psychiques* reported in 1919 that the French experienced more visions of death during the war. "Encore des visions de troupes aux combats," *Annales des sciences psychiques*, (1919): 63–65.

39. Charles Richet, "Lettre," p. 4.

40. Flammarion, *L'inconnu et les problèmes psychiques*.

41. William Reddy, *The Invisible Code: Honor and Sentiment in Postrevolutionary France, 1814–1848* (Berkeley: University of California Press, 1997); and Robert Nye, *Masculinity and the Male Codes of Honor in Modern France* (New York: Oxford University Press, 1993).

42. Richet, *Traité de métapsychique*, p. 72.

43. René Sudre, *Introduction à la métapsychique humaine* (Paris: Payot, 1926), p. 74.

44. Richet, *Traité de métapsychique*, p. 69.

45. Ibid., p. 758.

46. Boirac, *L'avenir des sciences psychiques*, pp. 52–53.

47. Maxwell, *Phénomènes psychiques*, p. 26.

48. Camille Flammarion, *Les maisons hantées; en marge de «La Mort et son mystère»* (Paris: E. Flammarion, 1923), p. 19.

49. Richet, *Traité de métapsychique*, pp. 63–68.

50. Ibid., p. 785.

51. E.g., Boirac, *L'Avenir des sciences psychiques*, p. 49.

52. Ibid., p. 34.

53. Paul Heuzé, *Où en est la métapsychique* (Paris: Gautier-Villars, 1926), p. 27.

54. Ibid., p. 46.

55. "Le Procès de la «Voyante de Saint-Quentin»," *Annales des sciences psychiques,* February 1906: 113.

56. "Le jugement dans le procès de la «Voyante de Saint-Quentin»," *Annales des sciences psychiques,* June 1906: 385.

57. Ibid., pp. 112–14; 385–87. Somnambulists were strictly forbidden to prescribe medication, but offering consultations was tolerated; see Georges-Denis Weil, *De l'exercice illégal de la médecine et de la pharmacie, législation pénale et jurisprudence* (Paris: Marchal & Billard, 1886), p. 49.

58. On Bompard's case and the medical debate over hypnotism during this period, see Ruth Harris, "Murder under Hypnosis in the Case of Gabrielle Bompard: Psychiatry in the Courtroom in Belle Epoque Paris," in *The Anatomy of Madness: Essays in the History of Psychiatry,* ed. W. F. Bynum, Roy Porter, and Michael Shepherd, vol. 2 (London: Tavistock Publications, 1985), pp. 197–241.

59. "L'envoûtement devant le tribunal," *Moniteur des études psychiques* 12 (1901): 287.

60. "Le magnétisme au Palais de justice," *Moniteur des études spirites* 10 (1901): 240.

61. "Un procès pour des manoeuvres spirites," *Annales des sciences psychiques,* 1909: 375. It is interesting to note that both of these cases involved women. There are many examples of court cases at the turn of the century in which women were presumed innocent on grounds of emotivity and released. See Edward Berenson, *The Trial of Madame Caillaux* (Berkeley: University of California Press, 1992); Ruth Harris, *Murder and Madness: Medicine, Law, and Society in the Fin de Siècle* (New York: Oxford University Press, 1989); Robert Nye, *Crime, Madness, and Politics in Modern France: The Medical Concept of National Decline* (Princeton, NJ: Princeton University Press, 1984); Ann-Louise Shapiro, *Breaking the Codes: Female Criminality in Fin-de-Siècle Paris* (Stanford: Stanford University Press, 1996).

62. The history of hauntings brought to the justice system was developed by the lawyer Fr. Zingaropoli in "Une maison hantée par les esprits. Droits du locataire à la résiliation du contrat de location," *Annales des sciences psychiques,* November 1907: 771–96; and by César de Vesme in "Les «Maisons hantées» devant la jurisprudence," *Revue métapsychique,* 1936: 94–113.

63. Maurice Maeterlinck, *The Unknown Guest* (New York: Dodd, Mead, 1914), pp. 378–82.

64. Flammarion, *Maisons hantées,* p. 96.

65. Ibid., p. 207.

66. Émile Tizané, *Il n'y a pas de maisons hantées? Journal d'un enquêteur incrédule, de 1925 à 1933* (Paris: Omnium littéraire, 1971). Eugène Osty, "La connaissance supra-normale et ses possibilités d'applications policières," *Revue métapsychique,* 1923: 81–84, relates how two mediums were consulted about a robbery and successfully identified the thief for the police.

67. Richet, *Traité de métapsychique,* pp. 739–40.

68. Ibid., pp. 776–80.

69. On animal wonders in the French context, see Sofie Lachapelle and Jenna Healey, "On Hans, Zou and the Others: Wonder Animals and the Question of Animal Intelligence in Early Twentieth-Century France" *Studies in History and Philosophy of Biological & Biomedical Sciences* 41 (2010): 12–20.

70. Maeterlinck, *Unknown Guest*, pp. 219–360.

71. Gifted animals were discussed on many occasions in the *Annales des sciences psychiques*. See, e.g., "Ce que sont devenus les chevaux d'Elberfeld et les chiens de Manheim durant la guerre," *Annales des sciences psychiques*, March 1916: 56–57; C. de Vesme, "Les chevaux pensants d'Elberfeld," ibid. 12 (December 1912): 353–63; Edmond Duchâtel, "Les animaux savants de Manheim," ibid. 10 (October 1913): 289–303; J. C. Ferrari, "Bêtes qui pensent," ibid., July 1912: 246–49; "Les chevaux d'Elberfeld et le 'Directeur de l'Institut de Psychologie zoologique,'" ibid., March 1914: 127–28; and William Mackenzie, "Les animaux savants de Manheim," ibid., February 1914: 40–51.

72. See M. A. Menegaux, "L'éducation des chevaux pensants d'Elberfeld," *Bulletin de l'Institut général psychologique* 3 (May–June 1913): 111–52; and Yves Delage, "Pour le contrôle des chevaux pensants d'Elberfeld," ibid.: 153–58.

73. Maeterlinck, *Unknown Guest*, pp. 249–54.

74. Ibid., pp. 269–78.

75. Ibid., pp. 305–7.

76. Ibid., pp. 336–44.

77. Duchâtel, *Annales des sciences psychiques*, 1913: 289–303.

78. Mackenzie, *Annales des sciences psychiques*, 1914: 40–53.

79. "Ce que sont devenus les chevaux d'Elberfeld et les chiens de Manheim durant la guerre," *Annales des sciences psychiques*, 1916: 56:

80. Richet, *Traité de métapsychique*, pp. 45–46.

81. Maxwell, *Phénomènes psychiques*, pp. 43, 315–16.

82. Ibid., p. 41.

83. Sudre, *Métapsychique humaine*, p. 51.

84. Albin Noiris, *Études sur le surnaturel-naturel. Première étude. Les tables tournantes et leurs dérives* (Lyon: Emm. Vitte, 1897), p. 4.

85. "L'Enquête de Jules Bois sur l'«Au-delà et les forces inconnues»," *Revue des études psychiques*, March–April 1902: 106.

86. Janet Oppenheim, *The Other World: Spiritualism and Psychical Research in England, 1850–1914* (Cambridge: Cambridge University Press, 1985) pp. 149–52.

87. Charles Blech on a séance with Eusapia at Monfort-l'Almaury in Guillaume de Fontenay, *À propos d'Eusapia Paladino. Les Séances de Monfort-l'Almaury, 25–28 juillet 1897): Compte rendu, photographies, témoignages et commentaires* (Paris: Société d'éditions scientifiques, 1898), p. 125.

88. Philippe de la Cortadière and Patrick Fuentes, *Camille Flammarion* (Paris: Flammarion, 1994).

89. Éugène Antoniadi, "Eusapia Paladino. Compte rendu de deux séances de spiritisme. Données chez M. C. Flammarion en Novembre 1898, et relatés par ÉUGÈNE

Antoniadi, astronome adjoint à l'Observatoire de Juvisy," Fonds Camille Flammarion de l'Observatoire de Juvisy-sur-Orge, p. 1.

90. Ibid., pp. 13–14.

91. Ibid., p. 11.

92. Ibid., p. 23.

93. Ibid., p. 5.

94. Ibid., p. 44.

95. Ibid., p. 39.

96. Ibid., pp. 16–17.

97. Fontenay, À propos d'Eusapia Paladino, p. 97.

98. Claude Langlois discusses the power of photography as a proof in "La photographie comme preuve entre médecine et religion," Histoire des sciences médicales 28 (1994): 325–36.

99. On the Buguet trial, see Hilarion-A.-B. Huguet (Dr), Spiritomanes et spiritophobes. Études sur le spiritisme, 5th ed. (Paris: l'auteur, 1875); and John Monroe, Laboratories of Faith: Mesmerism, Spirtism, and Occultism in Modern France (Ithaca, NY: Cornell University Press, 2008), pp. 150–98 (chap. 4). Some of Buguet's spirit photography can be seen in Clément Chéroux et al., The Perfect Medium: Photography and the Occult (New Haven, CT: Yale University Press, 2004), pp. 47, 56–59, 141.

100. Guillaume de Fontenay, La photographie et l'étude des phénomènes psychiques, abrégé de trois conférences données par l'auteur à la Société universelle d'études psychiques, en 1910 et 1911 (Paris: Cabaut, 1912), pp. 11–12.

101. "Pour la photographie transcendantale," Annales des sciences psychiques, September 1908: 307. For examples of discussions on spirit photography at the time, see Gabriel Delanne, "La photographie de la pensée," ibid., May 1908: 140–43; Guillaume de Fontenay, "Action de l'encre sur la plaque photographique," ibid., January 1909: 24–26; Commandant Darget, "Radio-activité humaine, rayons V (vitaux). Réponse à Fontenay," ibid.: 26–27; "Photographie de la pensée et des effluves humaines," ibid.: 20–24; "Un appareil pour photographier les esprits!" ibid., July 1912: 251–53.

102. Marcel Boll, Quelques sciences captivantes. Ondes humaines? Délires collectifs? Hypnotisme, psychanalyse, suggestion, métapsychie, astrologie, spiritisme, radiesthésie (Marseille: Éditions du Sagittaire, 1941), pp. 158–59.

103. William James and Théodore Flournoy, The Letters of William James and Theodore Flournoy, ed. Robert Charles Le Clair (Madison: University of Wisconsin Press, 1966), p. 227.

104. Richet, Traité de métapsychique, 656–64.

105. Ibid., pp. 125–26.

106. Heuzé, Où en est la métapsychique, p. 118.

107. Ibid., p. 120.

108. Ibid., p. 121.

109. Ibid.

110. Richet, Traité de métapsychique, p. 658.

111. Ibid., p. 113.

112. Heuzé, *Où en est la métapsychique*, p. 78.

113. Paul Heuzé, *Dernières histoires de fakirs* (Paris: Montaigne, 1932), p. 81.

CHAPTER 5: THE RISE AND FALL OF METAPSYCHICS

1. Rocco Santoliquido, letter of October 1918 (multiple copies), Archives de l'Institut métapsychique international, 3: 8.

2. Charles Richet, *Traité de métapsychique* (Paris: Alcan, 1922), p. 5.

3. Gustave Geley, *Comment faire progresser les études psychiques. Quelques voeux et projets* (Laval: L. Barnéoud, 1905), p. 3.

4. Geley himself provided the IMI with such an envelope. Attempts to identify his spirit after his death, however, were unsuccessful; see René Warcollier, "Rapport sur le test posthume du Dr. Gustave Geley," *Revue métapsychique* 4 (1956): 2–10.

5. Geley, *Comment faire progresser les études psychiques*, p. 10.

6. In 1947, the Office international d'hygiène publique was incorporated into the newly created World Health Organization.

7. Eugène Osty, "Le professeur R. Santoliquido," *Revue métapsychique* 6 (November–December 1930): 465–70.

8. Eugène Osty, "M. Jean Meyer," *Revue métapsychique* 2 (1931): 89–92.

9. Rocco Santoliquido, letter, October 1918, Archives de l'Institut métapsychique international, 3: 8.

10. Lodge, letter to Richet, October 16, 1918, ibid.; retranslated here from a French translation of the original English.

11. Ibid.

12. First session of the directing committee, Archives de l'Institut métapsychique international, 3: 8.

13. See the reverse of the title page in both the *Revue métapsychique* 1 (1921) and ibid., 1930.

14. Charles Richet, "Gustave Geley," *Revue métapsychique* 4 (July–August 1924): 271.

15. "Les conférences à l'Institut métapsychique," *Revue métapsychique* 1 (January–February 1926): 25; and ibid. 2 (March–April 1934): 144. Osty informed readers that he was available to answer questions of a metapsychical nature on Monday and Thursday afternoons.

16. René Sudre, "La Philosophie de Geley," *Revue métapsychique* 5 (September–October 1924): 344–46.

17. Gustave Geley, *L'ectoplasmie et la clairvoyance: Observations et experiences personelles* (Paris: Alcan, 1924) and Pierre Janet, *L'automatisme psychologique.* (Paris: Alcan, 1903), 401–2.

18. Gustave Geley, "Introduction à l'étude pratique de la médiumnité," *Revue métapsychique* 1 (January–February 1924): 58.

19. Charles Richet, "Gustave Geley," *Revue métapsychique* 4 (July–August 1924): 18.

20. Marja Wodzinska, "Apparition du docteur Geley à Varsovie," *Bulletin de la Société d'études psychiques de Nancy* 6 (May–June 1925): 85–87.

21. Eugène Osty, *La connaissance supra-normale*, 2nd ed. (Paris: Alcan, 1925), p. xix

22. Ibid., p. xxiv.

23. Eugène Osty and Marcel Osty, *Les pouvoirs inconnus de l'esprit sur la matière. Premières étapes d'une recherché* (Paris: Alcan, 1932).

24. Eugène Osty, "Projets et buts de recherches de l'Institut métapsychique international de Paris," *Revue métapsychique* 1 (January–February 1926): 8–9.

25. Ibid., 4.

26. Ibid., p. 24.

27. James McClenon, *Deviant Science: The Case of Parapsychology* (Philadelphia: University of Pennsylvania Press, 1984), makes a similar point about contemporary parapsychology by presenting a model explaining the stable position of parapsychologists. Parapsychologists want their discipline to be treated as scientific, but the funding of their work and their audience lie outside of the scientific world, making their survival depend on staying exactly where they are, at the margins.

28. René Sudre, letter to a member of the IMI (possibly René Warcollier), September 7, 1926, Archives de l'Institut métapsychique international, 18: 1.

29. René Sudre, *Introduction à la métapsychique humaine* (Paris: Payot, 1926), p. 10.

30. Daniel Berthelot, letter to the president, December 12, 1926, Archives de l'Institut métapsychique international, 22: 14.

31. Rocco Santoliquido, letter to Berthelot, December 15, 1926, ibid.

32. Garcon, letter to Léveillée, November 30, 1931, ibid., 22: 11.

33. Rocco Santoliquido, letter to Jean Meyer, January 15, 1929, ibid., 22: 14.

34. Rocco Santoliquido, letter to Jean Meyer, March 4, 1929, ibid.

35. Joseph Maxwell, letter to Osty, April 9, 1929, ibid., 22: 11.

36. Eugène Osty, letter to Reynaud, December 4, 1931, ibid.

37. Ibid., 22: 14.

38. Garcon, letter to Léveillée, November 30, 1931, ibid., 22: 11.

39. Garçon, letter to Léveillée, November 30, 1931, ibid.

40. Charles Richet quoted by Eugène Osty in "Charles Richet," *Revue métapsychique* 1 (January–February 1936): 30.

41. Osty recalled in his eulogy of Richet that the latter had planned to give his last lecture at the Sorbonne—a formal event, which many personalities would attend— on the subject of metapsychics. Uncomfortable at the prospect, the dean of the faculty reportedly asked Richet rather to discuss metapsychics in his penultimate lecture and to devote his last lecture to the work of the physiological laboratory of the faculty between 1881 and 1925. Richet diplomatically agreed to do so. Ibid., pp. 22–33.

42. Richet's *Traité de métapsychique* was presented, but his communication on the subject was not included in the Académie des sciences' *Comptes-rendus*. Ibid., p. 31.

43. Séance of January 3, 1911, *Comptes-rendus de l'Académie des sciences* 152 (1911): 17.

44. Conseil d'état, letter to Académie des sciences, February 11, 1905, Archives nationales, F17 13020.

45. Juliette de Reinach, letter to Académie des sciences, May 25, 1910, ibid.

46. Académie des sciences, acknowledgement to the Ministère de l'instruction publique of receipt of authorization to establish the Fanny Emden prize, July 27, 1910, ibid.

47. Séance of January 3, 1911, *Comptes-rendus de l'Académie des sciences* 152 (1911): 17.

48. "Le prix 'Fanny Emden' décerné par l'Académie française des sciences," *Annales des sciences psychiques,* November–December 1911: 370.

49. Séance of December 18, 1911, *Comptes-rendus de l'Académie des sciences* 153 (1911): 1378–80; and "Le mouvement psychique," *Annales des sciences psychiques,* November–December 1911: 369–70.

50. Séance of December 15, 1913, *Comptes-rendus de l'Académie des sciences* 157 (1913): 1297–1300; and *Annales des sciences psychiques,* 1913: 371.

51. Séance of December 2, 1918, *Comptes-rendus de l'Académie des sciences* 167 (1918): 865.

52. Séance of December 22, 1919, *Comptes-rendus de l'Académie des sciences* 169 (1919): 1285–86.

53. Séance of December 17, 1923, *Comptes-rendus de l'Académie des sciences.*

54. René Warcollier, "César de Vesme," *Revue métapsychique* 4–5 (July 1938): 242–43.

55. Séance of March 6, 1939, *Comptes-rendus de l'Académie des sciences* 697 (1939): 2087.

56. Paul Heuzé, *Où en est la métapsychique* (Paris: Gautier-Villars, 1926), p. 179.

57. For another account of these experiments, see Françoise Parrot, "Psychology Experiments: Spiritism at the Sorbonne," *Journal of the History of Behavioral Sciences* 29 (1993): 22–28.

58. On Henri Piéron, see Paul Fraisse, "French Origins of the Psychology of Behavior: The Contribution of Henri Piéron," *Journal of the History of Behavioral Sciences* 6 (1970): 111–19.

59. Heuzé, *Où en est la métapsychique,* 116–26. On the episode of the Villa Carmen, also see Blondel's essay in Bernadette Bensaude-Vincent and Christine Blondel, *Des savants face à l'occulte* (Paris: La Découverte, 2002).

60. Heuzé, *Où en est la métapsychique,* pp. 179–83.

61. Ibid., p. 331.

62. Juliette-Alexandre Bisson, *Les phénomènes de matérialisation. Étude expérimentale* (Paris: Alcan, 1914), pp. xiv–xv.

63. Heuzé, *Où en est la métapsychique,* p. 183.

64. Ibid., p. 196

65. Juliette-Alexandre Bisson, *Le médiumnisme et la Sorbonne* (Paris: Alcan, 1923), pp. 77–79.

66. Ibid., pp. 7–8.

67. Gustave Geley, "À propos des experiences de la Sorbonne," *Bulletin de la Société d'études psychiques de Nancy* 1 (1922): 29–30.

68. Gustave Geley, "À Propos des expériences de la Sorbonne," *Revue métapsychique* 4 (1922): 227.

69. Ibid., p. 230.

70. The report is reproduced in Heuzé, *Où en est la métapsychique*, pp. 198–202.

71. Ibid., pp. 203–36.

72. "Le mouvement psychique," *Annales des sciences psychiques*, 1904: 309.

73. "Discours prononcé le 28 mai 1913 par M. le professeur Henri Bergson," *Annales des sciences psychiques*, November–December 1913: 321.

74. *Compte-rendu officiel du premier congrès international des recherches psychiques à Copenhague* (Copenhagen: Éditeur scientifique, 1922), p. 11.

75. Ibid., pp. 528–29

76. René Sudre, "Compte-rendu du 2e congrès international de recherches psychiques," *Revue métapsychique* 5 (September–October 1923): 275.

77. Ibid., pp. 288–89.

78. *L'état actuel des recherches psychiques d'après les travaux du IIème Congrès international tenu à Varsovie en 1923 en l'honneur du Dr Julien Ochorewicz* (Paris: Presses universitaires de France, 1924), p. 21.

79. "Invitation au IIIe congrès international de recherches psychiques à Paris," *Revue métapsychique* 5 (1926): 379.

80. *Compte-rendu du IIIème congrès international de recherches psychiques à Paris* (Paris: Institut métapsychique international, 1928), p. 29.

81. Ibid.

82. Ibid., p. 31.

83. For a sympathetic account of the Paris congress, see A. Westermann, "Le IIIe congrès international de recherches psychiques," *Bulletin de la Société d'études psychiques de Nancy* 3 (November–December 1928): 128–31.

84. René Sudre, "Le congrès des recherches psychiques de Copenhague," *Revue métapsychique*, September–October 1921: 378.

85. Eugène Osty, "Le professeur R. Santoliquido," *Annales des sciences psychiques* 6 (November–December 1930): 469–70.

86. Ibid.

87. Ibid.

88. Eugène Osty, letter to Hans Driesch, April 3, 1928, Archives de l'Institut métapsychique international, 21: 12.

89. Ibid.

90. Ibid.

91. Eugène Osty, letter to Richet, April 3, 1928, ibid.

92. Presentation of the project sent to both Richet and Driesch, April 3, 1928, ibid.

93. Document on the foundation of the center, ibid.; and "Création à Genève d'un Centre permanent de conférences et de congrès internationaux de recherches psychiques," *Revue métapsychique* 3 (May–June 1928): 175–78.

94. Hans Driesch, letter to Osty, July 2, 1928, Archives de l'Institut métapsychique international, 21: 12.

95. Letter to Hans Driesch (not signed but probably from Osty), July 14, 1928, ibid.

96. Ibid.

97. Albert von Schrenck-Notzing, letter to Osty, July 19, 1928, ibid.

98. Albert von Schrenck-Notzing, letter to Osty, July 21, 1928, ibid.

99. Lodge and Richet, letter to Osty, July 23, 1928, ibid.

100. Albert von Schrenck-Notzing, letter to Osty, August 25, 1928, ibid.

101. Osty, letter to Albert von Schrenck-Notzing, September 1, 1928, ibid.

102. Osty, letter to Richet, September 19, 1938, ibid.

103. Osty, letter to Albert von Schrenck-Notzing, December 30, 1928, ibid.

104. Ibid.

105. Osty, letter to Montandon, November 24, 1930, ibid.

106. Montandon, letter to Osty, December 18, 1930, ibid.

107. *Transactions of the Fourth International Congress for Psychical Research*, ed. Theodore Besterman (London: Society for Psychical Research, 1930), p. 13. Besterman would later become an accomplished scholar whose work focused on the French Enlightenment and Voltaire particularly.

108. Albert von Schrenck-Notzing, letter to Osty, August 25, 1928, Archives de l'Institut métapsychique international, 21: 12.

109. *IIIème congrès international de recherches psychiques*, p. 27.

CONCLUSION

1. Cornelius Tabori, *My Occult Diary* (London: Rider, 1951), p. 62.

2. For more information on the IMI and its activities today, see www.metapsychique .org (accessed August 18, 2010).

3. E.g., *Bulletin de la parapsychologie*, 1955–58; *Cahiers métapsychiques, ésotériques et traditionnels*, 1950–54; *Conscience de vie. Organe de l'Institut d'études culturelles et psychiques*, 1953–74; *Paris-revue. Organe de la Société des sciences psychiques expérimentales de Paris*, 1932; *La recherche psychique. Revue mensuelle de l'Institut psychique lyonnais*, 1948–49; *La revue des sciences psychiques*, 1946; *La science métapsychique. Organe mensuel d'informations scientifiques du Centre de recherches de l'Institut des sciences métapsychiques appliqués*, 1954–55; *Les sciences métapsychiques*, 1946–47; and *La Tour Saint-Jacques*, 1955–58.

4. From 1955 until recently, the IMI's Paris address was 5 place Wagram.

5. See Seymour Mauskopf, *The Elusive Science: Origins of Experimental Psychical Research* (Baltimore: Johns Hopkins University Press, 1980); and James McClenon, *Deviant Science: The Case of Parapsychology* (Philadelphia: University of Pennsylvania Press, 1984).

6. "Un point d'histoire," *Revue métapsychique* 1 (1955): 52.

7. Ibid., pp. 52–53.

8. Ibid., p. 54.

9. René Warcollier, "Opinion d'un témoin," *Revue métapsychique* 1 (1955): 55–57.

10. René Pérot, "Souvenirs de Monsieur René Pérot sur Madame Bisson," *Revue métapsychique*, September 1968: 13.

11. There have been a few studies in psychology indicating that perception may be affected by expectation, values, and needs. See, e.g., J. S. Bruner and C. C. Goodman, "Value and Need as Organizing Factors in Perception," *Journal of Abnormal and*

Social Psychology 42 (1947): 33–44; Frédéric Gosselin, and Philippe G. Schyns, "Superstitious Perceptions Reveal Properties of Internal Representations," *Psychological Science* 14, no. 5 (2003): 505–9; and R. Dotsch, D. Wigboldus, O. Langner, and A. van Knippenberg, "Prejudiced People Have Biased Representations of Ethnic Faces," *Psychological Science* 19, no. 10 (2008): 978–80.

  12. René Sudre, *Introduction à la métapsychique humaine* (Paris: Payot, 1926), p. 70.
  13. Ibid.

Bibliographic Essay

There is a profusion of published sources on nineteenth- and early twentieth-century investigations of the supernatural, only a fraction of which can be mentioned in a short bibliographic essay. A great place to begin any research on the topic are the four volumes Louis Figuier's *Histoire du merveilleux* (Paris: Hachette, 1859–62), in which he described the many wonders of his time. On more specific subjects, the sources are numerous. For example, Guillard, *Table qui danse et table qui répond. Expériences à la portée de tout le monde* (Paris: Garnier frères, 1853), Ferdinand Silas, *Instruction explicative des tables tournantes* (Paris: Houssiaux & Dentu, 1853), and P.-F. Mathieu, *Un mot sur les tables parlantes suivi du crayon magique et du guéridon poète* (Paris: Jules Laisné, 1854), each provided early reports of the turning tables. Some of the religious condemnations of the phenomenon written at the time were Jules-Eudes de Mirville's *Pneumatologie. Des esprits et de leurs manifestations fluidiques*, 4th ed. (Paris: H. Vrayet de Surcy, 1858), André Pezzani's *Condamnation des manifestations spirites* (Paris: Ledoyen, 1860), Henri Carion's *Lettres sur l'évocation des esprits à Madame **** (Paris: Dentu, 1853), and Abbé Almignana's *Du somnambulisme, des tables tournantes et des médiums, considérés leurs rapports avec la théologie et la physique* (Paris: Dentu & Germer-Baillière, 1854).

Early scientific responses to the turning tables and other supernatural phenomena, such as the divining rod, include those by members of the Académie des sciences: Michel-Eugène Chevreul, *De la baguette divinatoire, du pendule explorateur et des tables tourantes, au point de vue de l'histoire, de la critique et de la méthode expérimentale* (Paris: Mallet-Bachelier, 1854), and Jacques Babinet, "Sciences des tables tournantes au point de vue de la mécanique et de la physiologie," *Revue des deux mondes* 5 (1854): 408–19, and "Les sciences occultes au XIXe siècle, les tables tournantes et les manifestations prétendus surnaturelles considérées au point de vue de la science de l'observation," ibid. 6 (1854): 510–33. Physicians also considered the turning tables and the séances, usually emphasizing their pathological nature. For example, Philibert Burlet, *Du spiritisme considéré comme cause d'aliénation mentale* (Lyon: Richard, 1863), and Paul Duhem, *Contribution à l'étude de la folie chez les spirites* (Paris: G. Steinheil, 1904), both discuss the dangers of the spiritist practice for mentally unstable mediums.

Within a few years of their arrival, the séances led several participants to suggest doctrines based on the phenomena. Most notably, Allan Kardec and his doctrine of spiritism are presented in *Le livre des esprits* (Paris: Didier, 1865), while *Le livre des médiums* (Paris: Librairie des sciences psychiques et spirites, n.d.) tells how to conduct a séance. Some other important works on spiritism include Léon Chevreuil, *On ne meurt pas* (Paris: Jouve, 1935); Gabriel Delanne, *Le phénomène spirites* (Paris: Leymarie, 1909), *Le spiritisme devant la science* (1885; Paris: Bibliothèque de philosophie spiritualiste, 1923), and *La réincarnation* (Paris: Vermet, 1985); and Léon Denis, *Après la mort* (Paris: Librairie des sciences psychiques, 1905).

The occultist movement was launched at mid-nineteenth century, when Eliphas Lévi wrote works including *Histoire de la magie* (Paris: Guy Tredaniel, 1996) and *La science des esprits* (Paris: Éditions de la Maisnie, 1976). Other prolific occultist writers included Papus (Dr. Gérard Encausse), with his numerous works on the topic, among them *La réincarnation* (1912; Paris: Dangles, 1953), *Traité élémentaire d'occultisme* (Paris: Diffusion scientifique, 1954), *Enseignement méthodique de l'occultisme* (Paris: Éditions de l'initiation, 1901–2), *La science des mages et ses applications théoriques et pratiques* (Paris: Librairie du merveilleux, Chamuel, 1892) and *La magie et l'hypnose* (Paris: Éditions traditionnelles, 1975). Papus hoped to develop a total science through his work. Others focused on more specific aspects of occultism, such as the human fluid. Some of the more important works on the topic were Hippolyte Baraduc's *Les vibrations de la vitalité humaine* (Paris: J.-B. Baillière et fils, 1904) and *La force curatrice à Lourdes et la psychologie du miracle* (Paris: Bloud, 1907), Guillaume de Fontenay's *La chimiographie et la prétendue photographie du rayonnement vital* (Paris: Société des publications scientifiques et industrielles, 1913), and Albert de Rochas' *L'extériorisation de la sensibilité* (Paris: Pygmalion, 1977) and *Les états profonds de l'hypnose* (Paris: Librairie générale des sciences occultes, Bibliothèque Chacornac, 1904).

Psychiatrists and psychologists shared an interest in the supernatural, usually attempting to explain it in pathological terms. Jean-Martin Charcot's classic article on the topic was "La foi qui guérit," *Revue hebdomadaire* (1892): 112–32. Other writers on faith and cures included both the convinced Catholic Antoine Imbert-Gourbeyre, author of *L'hypnotisme et la stigmatisation* (Paris: Bloud & Barral, 1899) and *La stigmatisation, l'extase divine et les miracles de Lourdes* (Clermont-Ferrand: Librairie catholique; Paris: Belle, 1894), and the ardent anticlericalist Désiré-Magloire Bourneville: *Science et miracle* (Paris: A. Delahaye, 1875) and his Bibliothèque diabolique series.

Based on his prolonged work with the medium Hélène Smith, Théodore Flournoy provided the most extensive study on mediumship by a psychologist in *Des Indes à la planète Mars. Étude sur un cas de somnambulisme avec glossolalie* (1899; Paris: Slatkine Reprints, 1983), translated by Daniel B. Vermilye as *From India to the Planet Mars: A Study of a Case of Somnambulism with Glossolalia* (New York: Harper & Bros., 1900; new ed., Princeton, NJ: Princeton University Press, 1994). Pierre Janet also studied the phenomena in *L'automatisme psychologique*, 4th ed. (Paris: Félix Alcan, 1903). The Institut général psychologique's report written on the séances with Eusapia Palladino can be found in the journal of the institute: Jules Courtier, "Rapport sur les séances d'Eusapia Palladino à l'Institut général psychologique en 1905, 1906, 1907 et 1908," *Bulletin de l'Institut général psychologique* 5–6 (1908).

On psychical research and metapsychics, the most relevant work to consult is probably Charles Richet's *Traité de métapsychique* (Paris: Félix Alcan, 1922). Other important works include Juliette-Alexandre Bisson, *Les phénomènes de matérialisation* (Paris: Félix Alcan, 1914) and *Le médiumnisme et la Sorbonne* (Paris: Alcan, 1923), Camille Flammarion's *Les maisons hantées* (Paris: E. Flammarion, 1923), *Les forces naturelles inconnues* (Paris: Didier, 1865), and *L'inconnu et les problèmes psychiques* (Paris: E. Flammarion, 1900), Gustave Geley, *L'ectoplasmie et la clairvoyance* (Paris: Félix Alcan, 1924), and *Introduction à l'étude pratique de la médiumnité* (Étampes: Terrier frères, 1924), Eugène Osty, *Connaissance supranormale* (Paris: Alcan, 1927), René Sudre, *Introduction à la métapsychique humaine* (Paris: Payot, 1926), and César de Vesme, *Histoire du spiritualisme expérimental* (Paris: Jean Meyer, 1928).

Proceedings of psychical research meetings in the 1920s are very useful to get a sense of the kind of work being accomplished in the field at the international level. They are *Compte rendu officiel du premier congrès international des recherches psychiques à Copenhague* (Copenhagen: Éditeur scientifique, 1922), *L'état actuel des recherches psychiques d'après les travaux du IIème Congrès international tenu à Varsovie en 1923* (Paris: Presses universitaires de France, 1924), *Compte rendu du IIIème congrès international de recherches psychiques à Paris* (Paris: Institut métapsychique international, 1928), and *Transactions of the Fourth International Congress for Psychical Research*, ed. Theodore Besterman (London: Society for Psychical Research, 1930). For a critical discussion of both psychical research and metapsychics, see Paul Heuzé, *Dernières histoires de fakirs* (Paris: Montaigne, 1932), *Fakirs, fumistes et cie* (Paris: Éditions de France, 1926), *Les morts vivent-ils?* (Paris: Renaissance du livre, 1921), *La plaisanterie des animaux calculateurs* (Paris: Éditions de France, 1928), and *Où en est la métapsychique* (Paris: Gautier-Villars, 1926).

Many journals directly related to various aspects of the investigation of the supernatural were published during the period. On spiritism the most important journal was Kardec's *Revue spirite* (1858–today). Others included the *Bulletin de l'Union spirite française* (1921–35), Pierre-Gaëtan Leymarie's *Bulletin mensuel de la Société scientifique d'études psychologiques* (1882–83), and Gabriel Delanne's *Revue scientifique et morale du spiritisme* (1896–1914, 1921–26) and *Le spiritisme* (1883–95). Occultist journals were numerous but tended to be short-lived. The most important ones were Papus's journal *L'initiation* (1888–1912) along with *La curiosité* (1889–98), the *Echo du monde occulte* (1905–6), the *Echo du merveilleux* (1897–1914), *L'étoile d'Orient* (1908–9), and *La haute-science* (1911). As for journals of theosophy in France, they included *L'aurore* (1886–95), *Le lotus* (1887–89), *Le lotus bleu* (1890–1940; 1946–86), and the *Revue théosophique* (1889–90).

The main journal of psychical research in France was the *Annales des sciences psychiques* (1890–1919), but there were others, mostly regional journals such as the *Bulletin de la Société d'études psychiques de Lyon* (1921–24), the *Bulletin de la Société d'études psychiques de Nancy* (1900–14, 1920–39), the *Bulletin de la Société d'études psychiques de Nice* (1912–14), the *Bulletin du Centre d'études psychiques de Marseille* (1902–13), *Le monde psychique* (1911–12), *Moniteur des études psychiques* (1901–3), *Le mouvement psychique* (1900–1901), *Revue des études psychiques* (1901–4), the *Revue générale des sciences psychiques* (1907–11), and *La tribune psychique* (1897–1984).

Metapsychics in France organized itself around the Institut métapsychique international, which still exists today (www.metapsychique.org [accessed August 18, 2010]). Its journal was the *Revue métapsychique* (1920–46; 1948–91). Throughout the 1940s and 1950s, new journals began to appear in France presenting themselves as either journals of psychical research, metapsychics, or parapsychology. Among them were the *Bulletin de la parapsychologie* (1955–58), *Cahiers métapsychiques, ésotériques et traditionnels* (1950–54), *Conscience de vie* (1953–74), *La recherche psychique. Revue mensuelle de l'Institut psychique lyonnais* (1948–49), *La revue des sciences psychiques* (1946), *La science métapsychique* (1954–55), and *La Tour Saint-Jacques* (1955–58).

Finally, there are some very useful archival sources on the topic, particularly at the Institut métapsychique international, where the papers from the early years of the organization and a good library are accessible. The Fonds Camille Flammarion situated in the astronomer's former home and observatory in Juvisy-sur-Orge (Île-de-France) contains a large quantity of material on just about anything Flammarion was interested in, including spiritism and psychical research. The Archives de l'Académie des sciences holds an envelope on the Fanny Emden Prize, as well as some information on former members of the academy. Some information on the spiritist movement including Allan Kardec's tour to promote spiritism in 1862 (F19 10927: Spirites et Kardec), congresses of psychology including the 1900 in Paris (F17 3091: Congrès divers en France et à l'étranger, 1833–1927), and the Fanny Emden Prize (F17 13020: Dons et legs à l'Académie des sciences, 1874–1925) can also be found in the Archives nationales in Paris.

ତ୍ଥ

There are a number of secondary sources related to the various aspects of the investigation of the supernatural in the nineteenth and twentieth century. Owen Chadwick's classic argument for the secularization of Europe during this period is found in *The Secularization of European Mind in the Nineteenth Century* (Cambridge: Cambridge University Press, 1975). For France, a revision has been undertaken and has shown the rich and varied religious experiences of the population of the period, most notably by Thomas Kselman in *Miracles and Prophecies in Nineteenth-Century France* (New Brunswick, NJ: Rutgers University Press, 1983) and *Death and the Afterlife in Modern France* (Princeton, NJ: Princeton University Press, 1993). Judith Devlin has shown that the supernatural continued to have a strong hold on the French population, particularly in the countryside in *The Superstitious Mind: French Peasants and the Supernatural in the Nineteenth Century* (New Haven, CT: Yale University Press, 1987).

Historical scholarship reflects the many ways in which the supernatural was concretized: Barbara Corrado Pope, "Immaculate and Powerful: The Marian Revival in the Nineteenth Century," in *Immaculate and Powerful: The Female in Sacred Image and Social Reality*, ed. Clarissa W. Atkinson, Constance H. Buchanan, and Margaret

R. Miles (Boston: Beacon Press, 1985), 173–200, and Sandra L. Zimdars-Swarty, *Encountering Mary. From La Salette to Medjugorje* (Princeton, NJ: Princeton University Press, 1991), have both, for example, explored the cultural significance of the numerous apparitions of the Virgin Mary during this period. Holy apparitions sometimes led to the building of shrines and the establishment of pilgrimage traditions, most notably that of Lourdes. Ruth Harris, *Lourdes: Body and Spirit in the Secular Age* (London: Allen Lane, Penguin Press, 1999), explores the development of both the story of the visions and the site, Suzanne K. Kaufman has focused on aspects of the pilgrimage in "Miracles, Medicine and the Spectacle of Lourdes: Popular Religion and Modernity in Fin-de-Siècle France" (Ph.D. diss. Rutgers University, 1996), and Jason Szabo has discussed the role of physicians at the sanctuary in "Seeing Is Believing? The Form and Substance of French Medical Debates over Lourdes," *Bulletin of the History of Medicine* 76 (2002): 199–230. Other claims of physical religious experiences during the period include the cases of demonic possessions in the small town of Morzine from 1857 to 1877 described in Jacqueline Carroy, *Le mal de Morzine. De la possession à l'hystérie (1857–1877)* (Paris: Solin, 1981), Ruth Harris, "Possession on the Borders: The 'Mal de Morzine' in Nineteenth-Century France," *Journal of Modern History* 69 (1997): 451–78, and Laurence Maire, *Les possédées de Morzine (1857–1873)* (Lyon: Presses universitaires de Lyon, 1981), and cases of stigmata such as that of Louise Lateau in Belgium, discussed in Sofie Lachapelle, "Between Miracle and Sickness: Louise Lateau and the Experience of Stigmata and Ecstasy," *Configurations* 12 (2004): 77–105.

Beyond the Catholic experience, the importance of the supernatural in a world becoming increasingly rational is apparent in the revival of an occultist tradition and the popularity of séances during the nineteenth century. This trend has been discussed in its various aspects. Most notably, in *Laboratories of Faith: Mesmerism, Spiritism, and Occultism in Modern France* (Ithaca, NY: Cornell University Press, 2008), John Warne Monroe has explored spiritism, mesmerism, and occultism and attempts to modernize faith by providing it with empirical evidence, and Lynn L. Sharp, *Secular Spirituality: Reincarnation and Spiritism in Nineteenth-Century France* (Lanham, MD: Lexington Books, 2006), has emphasized the social dimensions of the séances and the spiritist movement. In *Voyantes, guérisseuses et visionnaires en France, 1785–1914* (Paris: Albin Michel, 1995), Nicole Edelman has explored the world of female clairvoyants, healers, and visionaries in France, presenting mediums as part of a long lineage of female clairvoyants, and Bertrand Méheurst has discussed séances as part of his history of French animal magnetism in *Somnambulisme et médiumnité*, vol. 1, *Le défi du magnétisme* (Le Plessis–Robinson, Paris: Institut Synthélabo pour le progrès de la connaissance, 1999).

The literature on the séances that took place outside of France is now considerable, particularly in regards to the Anglo-Saxon context. Whereas in France séances rapidly centered on the spiritist doctrine, in Britain and in the United States they remained more fluid, and the movement was never as focused on specific teachings and conventions. Alex Owen, *The Darkened Room: Women, Power and Spiritualism in Late Victorian England* (Philadelphia: University of Pennsylvania Press, 1990), and Ann D. Braude, *Radical Spirits: Spiritualism and Women's Rights in Nineteenth Century*

America (Boston: Beacon Press, 1989), have discussed the development of female mediumship as a source of empowerment. Logie Barrow has presented the séances as sites of protest, focusing on the plebeian culture associated with them in *Independent Spirits: Spiritualism and English Plebeians, 1850–1910* (New York: Routledge & Kegan Paul, 1986). Ruth Brandon has written a history of the séances focusing on fraud in *The Spiritualists: The Passion for the Occult in the Nineteenth and Twentieth Centuries* (New York: Knopf, 1983). More recently, Barbara Weisberg has written the history of the early American movement focusing particularly on the Fox sisters in *Talking to the Dead: Kate and Maggie Fox and the Rise of Spiritualism* (New York: Harper Collins, 2004).

On the French occultist movement, David Allen Harvey has explored the history of the Martinist movement founded by Papus and Stanilas de Guaita in *Beyond Enlightenment: Occultism and Politics in Modern France* (DeKalb: Northern Illinois University Press, 2005), while Christopher McIntosh has written on Eliphas Lévi and the midcentury occult revival in *Eliphas Lévi and the French Occult Revival* (London: Rider, 1972). On the more artistic aspects of the movement, see Robert Pincus-Witten, *Occult Symbolism in France: Joséphin Péladan and the Salons de la Rose-Croix* (New York: Garland, 1976). Outside of France, Alex Owen, *The Place of Enchantment: British Occultism and the Culture of the Modern* (Chicago: University of Chicago Press, 2004), has presented British occultism in the context of an emergent modernity. More generally, James Webb, *The Occult Establishment* (LaSalle, IL: Library Press / Open Court, 1976) has discussed the importance of occultism in modern times and Sylvia L. Cranston has provided a useful bibliography of Blavatsky and a history of the Theosophical Society in *The Extraordinary Life and Influence of Helena Blavatsky, Founder of the Modern Theosophical Movement* (New York: Putnam, 1993).

Psychiatry and psychology have both historically had strong links to study of the supernatural. A growing number of historians have written or edited works on the topic; see particularly *Des savants face à l'occulte* (Paris: La Découverte, 2002), ed. Bernadette Bensaude-Vincent and Christine Blondel; Jacqueline Carroy, *Hypnose, suggestion et psychologie. L'invention de sujets* (Paris: Presses universitaires de France, 1991); Adam Crabtree, *From Mesmer to Freud: A Magnetic Sleep and the Roots of Psychological Healing* (New Haven, CT: Yale University Press, 1993); Sarah Ferber, "Charcot's Demons. Retrospective Medicine and Historical Diagnosis in the Writings of the Salpêtrière School," in *Illness and Healing Alternatives in Western Europe*, ed. Marijke Gijswijt-Hofstra, Hilary Marland, and Hans de Waardt (New York: Routledge, 1997), pp. 120–40; Ian Hacking, *Rewriting the Soul: Multiple Personality and the Sciences of Memory* (Princeton, NJ: Princeton University Press, 1984); Ruth Harris, "The Unconscious and Catholicism in France," *Historical Journal* 47 (2004): 331–54; Wilma Koutstaal, "Skirting the Abyss: A History of Experimental Exploration of Automatic Writing in Psychology," *Journal of the History of the Behavioral Sciences* 28 (1992): 5–27; Pascal Le Maléfan, *Folie et spiritisme. Histoire du discours psychopathologique sur la pratique du spiritisme ses abords et ses avatars (1850–1950)* (Paris: L'Harmattan, 1999); Mark S. Micale, *Approaching Hysteria: Disease and Its Interpretations* (Princeton, NJ: Princeton University Press, 1994), pp. 260–84; and Régine Plas, *Naissance d'une science humaine, la psychologie. Les psychologues et le merveilleux psychique* (Rennes: Presses universitaires de Rennes, 2000).

More generally, there have been many works on the history of nineteenth-century French psychiatry, among them Henri Baruk, *La psychiatrie française de Pinel à nos jours* (Paris: Presses universitaires de France, 1976); Robert Castel, *L'ordre psychiatrique* (Paris: Minuit, 1976); Ian R. Dowbiggin, *Inheriting Madness: Professionalization and Psychiatric Knowledge in Nineteenth-Century France* (Berkeley: University of California Press, 1991); Henri F. Ellenberger, *The Discovery of the Unconscious: The History and Evolution of Dynamic Psychiatry* (New York: Basic Books, 1970); and Jan Goldstein, *Console and Classify: The French Psychiatric Profession in the Nineteenth Century* (Cambridge: Cambridge University Press, 1987).

On the history of French psychology, see, in general, John I. Brooks III, *The Eclectic Legacy: Academic Philosophy and the Human Sciences in Nineteenth-Century France* (Newark, NJ: University of Delaware Press, 1998), and Jacqueline Carroy, Annick Ohayon, and Régine Plas, *Histoire de la psychologie en France, XIXe–Xxe siècles* (Paris: La Découverte, 2006). And on the beginnings of an experimental tradition in French psychology in the 1890s, see Jacqueline Carroy and Regine Plas, "The Origins of French Experimental Psychology: Experiment and Experimentalism," *History of the Human Sciences* 9 (1996): 73–84; Laurent Mucchielli, "Aux origines de la psychologie universitaire en France (1870–1900). Enjeux intellectuals, contexte politique et stratégies d'alliance autour de la *Revue philosophique* de Théodule Ribot," *Annals of Science* 55 (1998): 263–89; Serge Nicolas, Juan Segui, and Ludovic Ferrand, "*L'Année psychologique*: History of the Founding of a 100-Year-Old French Journal," *History of Psychology* 3 (2000): 44–61.

Little has been written on the history of psychical research and metapsychics in France. The IMI has been mentioned in a few histories of parapsychology, most notably in the sympathetic accounts of Robert Amadou, *La parapsychologie* (Paris: Denoël, 1954); and Robert Tocquet, *Les pouvoirs secrets de l'homme* (Paris: Productions de Paris, 1963). More recently, Bertrand Méheust in has written a history of magnetism and psychical research in France in *Somnambulisme et médiumnité*, vol. 2: *Le choc des sciences psychiques* (Le Plessis–Robinson, Paris: Institut Synthélabo pour le progrès de la connaissance, 1999), and Françoise Parrot, "Psychology Experiments: Spiritism at the Sorbonne," *Journal of the History of Behavioral Sciences* 29 (1993): 22–28, has discussed Eva C.'s séances at the Sorbonne in 1922.

Outside of France, British psychical research has attracted most attention. In particular, John J. Cerullo, *The Secularization of the Soul: Psychical Research in Modern Britain* (Philadelphia: Institute for the Study of Human Issues, 1982), and Janet Oppenheim, *The Other World: Spiritualism and Psychical Research in England, 1850–1914* (Cambridge: Cambridge University Press, 1985), have discussed the history of psychical research and the Society for Psychical Research (SPR). For an internal history of the SPR, see Renee Haynes, *The Society for Psychical Research, 1882–1982: A History* (London: Macdonald, 1982). Laurence R. Moore, *In Search of White Crows: Spiritualism, Parapsychology, and American Culture* (New York: Oxford University Press, 1977), has written on American psychical research, while the German side of the story has been told by Corinna Treitel, *A Science for the Soul: Occultism and the Genesis of the German Modern* (Baltimore: John Hopkins University Press, 2004), and Heather Wolffram, *The Stepchildren of Science: Psychical Research and Parapsychology in Germany, c. 1870–1939* (New York: Rodopi, 2009). On the more recent history of parapsychology and its origins, see, e.g., Brian Inglis, *Natural and Supernatural* (London: Hodder & Stoughton, 1977) and *Science and Parascience: A History of the Paranormal, 1914–1939* (London: Hodder & Stoughton, 1984), and Seymour Mauskopf, *The Elusive Science: Origins of Experimental Psychical Research* (Baltimore: Johns Hopkins University Press, 1980).

The various investigations of the supernatural occurred in the context of both the professionalization and the popularization of science in France. Maurice Crosland has written numerous books and articles in which he describes the processes by which science functioned as a profession throughout the nineteenth century, among them "Assessment by Peers," *Minerva* 24 (1986): 413–32; "The Emergence of Research Grants," *Social Studies of Science* 19 (1989): 71–100; "Scientific Credentials," *Minerva* 19 (1981): 605–31; and *Science Under Control: The French Academy of Sciences, 1795–1914* (Cambridge: Cambridge University Press, 1992). More specifically, John L. Davis has argued that through contests and prizes, the Académie des sciences was able to dominate electrical innovations in France, imposing scientific acceptance to the technological process in "Artisans and Savants: The Role of the Academy of Sciences in the Process of Electrical Innovation in France, 1850–1880," *Annals of Science* 55 (1998): 291–314, and Mary Jo Nye has discussed the limitations of the organization of science in France in "N-Rays: An Episode in the History and Psychology of Science," *Historical Studies in the Physical Sciences* 11 (1980): 125–56. Dorinda Outram has explored the role of space in the professionalization of science in France, arguing that science left the private sphere of the scientist's home for the public space of the museum and the laboratory during the nineteenth century in "New Spaces in Natural History," in *Cultures of Natural History*, ed. N. Jardine, J. A. Secord, and E. C. Spary (Cambridge: Cambridge University Press, 1996), pp. 249–65. Finally, *La Science pour tous: Sur la vulgarisation scientifique en France de 1850 à 1914*, ed. Bruno Béguet (Paris: Bibliothèque du conservatoire national des arts et métiers, 1990), and *La science populaire dans la presse et l'édition, XIXe et XXe siècles*, ed. Bernadette Bensaude-Vincent and Anne Ramussen (Paris: CNRS éditions, 1997) discuss the popularization of science in France in the nineteenth and twentieth centuries. Additional sources on specific topics can be found in the notes to each chapter.

Index